LANDSCAPE
Ecology and Agroecosystems

R.G.H. Bunce
L. Ryszkowski
M.G. Paoletti

LEWIS PUBLISHERS
Boca Raton   London   New York   Washington, D.C.

thanks to the financial support of the European Economic Commission (EEC) and the Italian CNR.

Landscape structure is a result of natural processes interfaced with voluntary and unconscious actions by man. Hence a system analysis of the countryside cannot be restricted to only one category of driving forces. A modern analysis of the landscape should deal with all the recognized processes, including their mutual interrelationship. Such an analysis constitutes a challenge for contemporary science and this book attempts to provide the reader with an insight into this very subject.

Robert G. H. Bunce
Grange-over-Sands

Lech Ryszkowski
Poznan

Maurizio G. Paoletti
Padova

About the Editors

Robert G. H. Bunce is a Research Ecologist with the Institute of Terrestrial Ecology and has been based at Merlewood Research Station in the Lake District, northwest England, for 27 years. He graduated from University College of North Wales, Bangor, where he also completed his Ph.D. on the ecology of mountain cliff vegetation. He has been concerned with the development of objective methods of vegetation and environmental and subsequent monitoring — initially in woodlands but now throughout all habitats in Britain. Environmental strata have been defined for the whole of Britain and used for sequential sampling surveys in 1978, 1984, and 1990 of land use, vegetation, and landscape features. These data have tracked the changes in the British countryside and have also been used for modeling exercises, e.g., estimating wood energy and wind potential and for impact assessment. Over the last few years, Dr. Bunce has worked largely under contract to various government agencies to provide ecological information on policy issues. He has written more than 50 contract reports in addition to over 100 articles both in the scientific and popular press, as well as editing several books.

Dr. Bunce has attended all the major meetings on landscape ecology in Europe and is chairman of the newly formed British section of the International Association. He also lectures at universities throughout Britain and Europe. Dr. Bunce is committed to developing cooperation between ecololgists at the European level and to increasing environmental awareness, especially in mountaineering circles.

Lech Ryszkowski is Professor of Natural Sciences and is Director of the Research Centre for Agricultural and Forest Environment of the Polish Academy of Sciences in Poznań, Poland. He graduated from the Faculty of Biology and Earth Sciences in 1955 and received his Doctorate from Warsaw University in 1963. He became an Associate Professor in 1971 and a Full Professor in 1985.

Dr. Ryszkowski is a member of IUCN-Ecological Section; INTECOL, of which he is head of the International Group for Agrecosystems; SCOPE; MaE; and the Commission for Rural Areas, Nature Protection, and Landscapes of EC. In Poland he is active in the National Council of Natural Resource Protection, the Ecological Committee of the Polish Academy of Sciences, and the Ecological Council of the President.

Dr. Ryszkowski's major scientific interests are in agroecology, environment protection and ecological guidelines of the protection of the agricultural landscape based on energy flow and matter cycling principles. Particular significance is emphasized on the possibilities of control on nonpoint pollution sources by the formation of biogeochemical barriers such as woodlots, shelterbelts, grasslands, water bodies and other, differentiating agricultural landscape. By these means the biological diversity of rural areas is maintained which is of a great importance for the strategy of living resource preservation. Dr. Ryszkowski has published over 180 papers and 16 books and organized numerous scientific conferences and symposia.

Maurizio G. Paoletti is a Soil Zoologist and Ecologist and has been a full-time Researcher in the Department of Biology, Padova University, since 1983. He received a degree in Natural Sciences from Padova University in 1973. In 1978 he received a Masters degree in Human Ecology from Padova University, and in 1978 a masters degree in Biological Control from the University of California, Berkeley, U.S.A.

Dr. Paoletti teaches entomology, zoology, soil ecology and agroecology at Padova University and since 1991 he has participated in two EEC-TEMPUS Masters' courses held in Budapest, Hungary and Sofia, Bulgaria on the subject of agroecology. Recently he spent sabbaticals at Ohio State University (1987) and Cornell University, N.Y. (1992) as a visiting professor. In addition, he has been involved with teaching at the Universities of Innsbruck, Austria; Torun, Poland; and Zaragoza, Spain.

Dr. Paoletti's current interests include cave and soil invertebrates, zoology, general ecology and agroecology. Biodiversity and interactions in natural and cultivated ecosystems have also attracted his attention. He has published over 125 papers and edited 12 volumes in the field.

Chapter Authors

J. ALVAREZ,
    Instituto Pirenaico de Ecologia, CSIC, Zaragoza, Spain

G. BALENT,
    Institut National de la Recherche Agronomique, Unite de Recherche sur les Systèmes Agraires et le Développement, Toulouse, France

A. BANACH,
    Department of Ecology, Warsaw University, Warsaw, Poland

J. BAUDRY,
    Institut National de la Recherche Agronomique, Départment Systèmes Agraires et Developpement, Lieury, France

A. BELLO,
    Cantro de Ciencias Medioambientales CSIC, Serrano, Madrid, Spain

R. G. H. BUNCE,
    Institute of Terrestrial Ecology, Merlewood Research Station, Grange-over Sands, Cumbria, United Kingdom

F. BUREL,
    Centre National de la Recherche Scientifique, Université de Rennes 1, Laboratoire d'Evolution des Systemes Naturels et Modifiés, Rennes, France

S. E. CARTER,
    Agroecological Studies Unit, Centro Internacional de Agricultura Tropical, Cali, Colombia

F. CASTIONI,
    Dipartimento di Agronomia e Produzioni Erbacee, Università degli Studi di Firenze, Italy

G. CERRETELLI,
    Dipartimento di Agronomia e Produzioni Erbacee, Università degli Studi di Firenze, Italy

P. CORONA,
    Istituto di Assestamento e Tecnologia forestale, Università di Firenze, Firenze, Italy

L. Di Benedetto,
  Instituto di Biologia e Ecologicia Vegetale, Universita di Catania, Catania, Italy

A. Di Meo,
  Dipartimento di Agronomia e Produzioni Erbacee, Università degli Studi di Firenze, Italy

K. Dobrowolski,
  Institute of Ecology, Polish Academy of Sciences, and Department of Ecology, Warsaw University, Warsaw, Poland

O. Failla,
  Instituto Coltivazioni Arboree, Università di Milano, Milano, Italy

A. Farina,
  Comune Di Aulla, Mueso di Storia Naturale della Lunigiana, Aulla, Italy

L. G. Firbank,
  Institute of Terrestrial Ecology, Monks Wood, Abbots Ripton, Huntingdon, Cambridgeshire, United Kingdom

A. Gómez-Sal,
  Instituto Pirenaico de Ecologia, CSIC, Zaragoaza, Spain

R. Groppali,
  Istituto di Entomologia, Pavia, Italia

H. Gulinck,
  Institute for Land and Water Management, Catholic University Leuven, Leuven, Belgium

C. J. Hallam,
  Institute of Terrestrial Ecology, Merlewood Research Station, Grange-over Sands, Cumbria, United Kingdom

P. G. Jones,
  Agroecological Studies Unit, Centro Internacional de Agricultura Tropical, Cali, Colombia

R. H. G. Jongman,
  Department of Physical Planning, Agricultural University, Wageningen, The Netherlands

J. T. R. Kalkoven,
  Research Institute for Nature Management, Department of Landscape Ecology, Leersum, The Neitherlands

J. Karg,
  Research Center for Agricultural and Forest Environment Studies, Polish Academey of Sciences, Poznan, Poland

A. KOZAKIEWICZ,
   Department of Ecology, Warsaw University, Warsaw, Poland

M. KOZAKIEWICZ,
   Department of Ecology, Warsaw University, Warsaw, Poland

J. C. LEFEUVRE,
   Centre National de la Recherche Scientifique, Universite de Rennes 1, Laboratoire d'Evolution des Systemes Naturels et Modifiés, Rennes, France

M. H. LOSVIK,
   Sogn og Fjordane College, Sogndal, Norway

F. LUCIANI,
   Cattedra di Botanica, Università di Catania, Catania, Italy

G. MARGARIT,
   Research Institute for Plant Protection, Bucharest, Romania

V. MARTIN MARTIN,
   Departamento de Geografia, Universidad de La Laguna, Canary Islands

G. MAUGERI,
   Cattedra di Botanica, Università di Catania, Catania, Italy

M. A. MUÑOZ-YANGUAS,
   Istituto Pirenaico de Ecologia, CSIC, Zaragoza, Spain

D. NOTA,
   Regione Toscana, Florence, Italy

S. PADERI,
   Regione Toscana, Florence, Italy

M. G. PAOLETTI,
   Department of Biology, University of Padova, Padova, Italy

F. PAUWELS,
   Institute for Land and Forest Management, Catholic University Leuven, Leuven, Belgium

E. POLI MARCHESE,
   Cattedra di Botanica, Università di Catania, Catania, Italy

S. RAZZARA,
   Catedra di Botanica, Università di Catania, Catania, Italy

S. REBOLLO,
   Servicio de Investigaciónes Agrarias, Apdo Oficial, Salamanca, Spain

M. RIGHINI,
   Dipartimento di Agronómia e Produzioni Erbacee, Università degli Studi Firenze, Firenze, Italy

D. M. ROBISON,
   Agroecological Studies Unit, Centro Internacional de Agricultura Tropical, Cali, Colombia

W. RODRIGUEZ BRITO,
   Departamento de Geografia, Universidad de La Laguna, Canary Islands, Spain

L. RYSZKOWSKI,
   Research Center for Agricultural and Forest Environment Studies, Polish Academy of Sciences, Poznan, Poland

N. SAUGET,
   Institut National de la Recherche Agronomique, Unite de Recherche sur les Systèmes Agraires et le Développement, Toulouse, France

A. SCIENZA,
   Istituto di Coltivazioni Arboree, Università di Milano, Milano, and Istituto Agrario Provinciale, S. Michele, Italy

G. TARTONI,
   Dipartimento di Agronomia e Produzioni Erbacee, Università degli Studi Firenze, Firenze, Italy

L. VALENTI,
   Istituto Coltivazioni Arboree, Università di Milano, Milano, Italy

C. VAZZANA,
   Dipartimento di Agronomia e Produzioni Erbacee, Università degli Studi di Firenze, Firenze, Italy

R. ZLOTIN,
   Institute of Geography AS U.S.S.R., Moscow, Russia

Contents

Preface iii

About the Editors v

Chapter Authors viii

Contents xii

PART 1 LANDSCAPE ECOLOGY

1. An Introduction to Landscape Ecology
   by R. G. H. Bunce and R. H. G. Jongman 3

2. The Ecological Significance of Linear Features in
   Agricultural Landscapes in Britain
   by R. G. H. Bunce and C. J. Hallam 11

3. Landscape Dynamics and Farming Systems: Problems of
   Relating Patterns and Predicting Ecological Changes
   by J. Baudry 21

4. Landscape Structure and the Control of Water Runoff
   by F. Burel, J. Baudry, and J. C. Lefeuvre 41

5. Agricultural Transport and Landscape Ecological Patterns
   by F. Pauwels and H. Gulinck 49

6. Effect of Habitat Barriers on Animal Populations and
   Communities in Heterogenous Landscapes
   by K. Dobrowolski, A. Banach, A. Kozakiewicz, and
   M. Kozakiewicz 61

7. Above Ground Insect Biomass in Agricultural Landscapes
   of Europe
   by L. Ryszkowski, J. Karg, G. Margalit, M. G. Paoletti,
   and R. Zlotin 71

8. Survival of Populations and the Scale of the Fragmented,
   Agricultural Landscape
   by J. T. R. Kalkhoven 83

9. Implications of Scale on the Ecology and Management of Weeds
   by L. G. Firbank 91

10. Total Species Number as a Criterion for Conservation of Hay Meadows
    by M. H. Losvik 105

11. The Diversity of Agricultural Practices and Landscape Dynamics: The Case of a Hill Region in the South West of France
    by N. Sauget and G. Balent 113

12. The Role of Natural Vegetation in the Agricultural Landscape for Biological Conservation in Sicily
    by L. Di Benedetto, F. Luciani, G. Maugeri, E. Poli, and S. Razzara 131

PART II AGROECOSYSTEMS

13. Patterns of Change in the Agrarian Landscape in an Area of the Cantabrian Mountains (Spain). Assessment by Transition Probabilities
    by A. Gomez-Sal, J. Alvarez Martionez, M. A. Muñoz-Yanguas, and S. Rebollo 141

14. Breeding Birds in Traditional Tree Rows and Hedges in the Central Po Valley (Province of Cremona, Northern Italy)
    by R. Groppali 153

15. Bird Fauna in the Changing Agricultural Landscape
    by A. Farina 159

16. Study Outline on Ecological Methods of Afforestation
    by P. Corona 169

17. Research on Germplasm of Herbaceous Plants in Tuscany
    by F. Castioni, G. Cerretelli, A. Di Meo, D. Nota, S. Paderi, M. Righini, G. Tartoni, and C. Vazzana 177

18. Grapevine Germplasm Diversity and Conservation
    by A. Scienza, O. Failla, and L. Valenti 183

19. A Geographical Information Approach for Stratifying Tropical Latin America to Identify Research Problems and Opportunities in Sustainable Agriculture
    by D. M. Robison, P. G. Jones, and S. E. Carter 197

20. Ecological Aspects of Production in the Canary Islands Traditional Agrosystems
    by V. Martin Martin, W. Rodriguez Brito, and A. Bello 215

Index 231

PART I
Landscape Ecology

1 An Introduction to Landscape Ecology

R. G. H. BUNCE
R. H. G. JONGMAN

Abstract: *Landscape ecology is an interdisciplinary field of science that may be defined as the study of the interaction between spatial and temporal components in a landscape and its associated flora and fauna. Underlying this theme is the realization that a landscape is more than the sum of its parts because of the interdependence of species between the various elements.*

This paper first describes the history of landscape ecology, before summarizing the principal concepts involved and possible directions for future developments.

INTRODUCTION

Landscape ecology originally developed on the interface between physical geography and ecology. In German literature the term was first introduced by Troll (1939) for the inclusion of ecology in physical geography, while in English literature Tansley (1935) introduced the concept of the ecosystem, although the latter is more concerned with vegetation patterns. In Russian (Sukachev and Dylis, 1964), Czechoslovakian, and Dutch research (Vink, 1983; Zonneveld, 1972), and later in American and French research (Risser et al., 1983; Forman and Gordon, 1986), the current discipline of landscape ecology was developed.

Some aspects of landscape ecology overlap other older disciplines, e.g., autecology and synecology, which strictly should only be termed landscape ecology if they are considered at a landscape level and have appropriate objectives (Jongman et al., 1987). The dominant role of ecology is important in the application of landscape ecology practices that ameliorate continued degradation of the environment. As such, the history of the landscape and knowledge of past management procedures are important topics, as they assist man in his ability to control the system that makes up a landscape.

It is no coincidence that the highly populated countries of The Netherlands and Belgium have led the way in landscape ecology, because the pressures on the landscape have led scientists to concentrate on the development of the study of interactions between landscape elements and the way they are managed and used by man.

It is also significant that many eastern European countries, because of their traditional involvement in landscape, have developed major programs of landscape ecology. Here, the traditional starting point is physical geography and the emphasis is more on pattern analysis and planning (Ruzicka and Miklos, 1981; Miklos, 1989) and landscape theory (Neef, 1981) than on process analysis.

HISTORY

The initial concept of landscape ecology was conceived by Troll in 1935, who later developed further concepts in the following years. During the last decades, the conceptual trend moved toward integrated analysis of landscape processes and patterns. The first international meeting of landscape ecologists was held in 1981 in Veldhoven, The Netherlands, with scientists from many countries. The International Association for Landscape Ecology (IALE) was founded in October, 1982, in Piestany (CSSR).

In both meetings many of the papers submitted were more like theoretical concepts rather than scientific results, which helped develop a new discipline concerning integration at the landscape level. However, the next symposium, held in 1984 in Roskilde, shaped the development of a core of science that involved different aspects of ecology but concentrated on the landscape level rather than the patch level.

A further difference with traditional ecological papers is that there is an emphasis upon the manipulation of the environment and relationships between the elements of the vegetation, the fauna, the soils, and of the agricultural ecosystems that emphasize the distinctive nature of the ecology. This was further emphasized in 1987 at the symposium in Münster on connectivity in landscapes, with emphasis being on the interdependence of elements in the landscape that are fundamental to landscape ecology (Schreiber, 1988). The papers at this meeting concentrated consistently on the way that flora and fauna are linked to elements within the landscape and upon the diversity of these linkages. This symposium set a trend for the development of landscape ecology. In due course, it was followed by the first issue of the *Journal for Landscape Ecology.*

Various key papers over the years have provided a good introduction to the subject of landscape ecology. These papers are still useful for readers not familiar with the subject and can be found in the references given at the end of this chapter. Recent books by Naveh and Lieberman (1984), Forman and Gordon (1986), and Zonneveld and Forman (1990) are particularly useful. The first describes the theoretical basis behind landscape ecology and gives a good overall introduction to the topic. Forman and Gordon (1986) introduce landscape ecology at a more basic level, while Zonneveld and Forman (1990) offer an overview on the recent rapid developments in landscape ecological research and their importance for management and planning.

Urban et al. (1987) provide an important text on landscape ecology, as they discuss the relationships between elements within the landscape and their interdependence as well as the role of man in their management and manipulation. Opdam (1987, 1988), Van Dorp and Opdam (1987), and Harms and Opdam (1990) have shown the key importance of concepts relating individual populations in the landscape in a biogeographical sense and their relationship with the wider countryside. Such ideas originated from principles in biogeography and its original theoretical distinctions, but have been developed to involve fragmentation and isolation of populations in the landscape at a practical level. In this context, Harris (1984) describes an elegant way to apply island theory in forestry planning and management.

Sharp and Keddy (1986) show the importance of the isolation of populations and the way in which, historically, certain elements in the landscape have actually become isolated from each other. This theme has been further developed in the work of Baudry (1984) and Asselin (1988) for spiders, and Burel (1989) for carabid beetles. Experimental research on the effects of isolation of arthropods has been carried out in agricultural landscapes by Mader et al. (1986). This comparison between geographical isolation and the linking of dispersion between animal and botanical elements is fundamental to landscape analysis. Harms and Knaapen (1988), in a paper on squirrels, show how management can be modelled and, with knowledge of the dispersal patterns of an organism, how the landscape can be manipulated so that the animals can disperse throughout islands which are not too far apart.

Fragmentation of the landscape is frequently covered by workers in other fields, but is not always labeled as landscape ecology. Thus, Webb (1986), in his classic book on the British lowland heaths, describes the progressive fragmentation of landscapes involving the heathlands of Dorset (using many of the theorems of landscape ecology). Although this is a basic text describing the loss of habitat to agriculture, the effects are important at a landscape (ecology) level because of the implications for the ecological web of the fauna (Andrewartha and Birch, 1984) and for integration between the elements of the landscape. Another classic and, for landscape ecology, basic study was done by Den Boer (1971, 1977) on the dynamics of the carabid populations in various landscapes. He showed that before landscape ecology was amended to include the concept of connectivity, landscape connectivity is necessary for most populations to survive.

Further work by Burel (1988), Burel and Baudry (1990), Verboom and Van Apeldoorn (1990), among others, has shown that the various individual elements within a landscape are dependent upon each other. The pathways followed by the species are determined by the mosaics present. Classic autecological factors which control the behavior of an individual species can determine its subsequent pattern of distribution throughout the landscape. For example, an aggressive plant species requiring high nitrogen levels can spread rapidly through abandoned agricultural land. Ecologists traditionally are used to dealing with landscapes in a relatively stable equilibrium. However, modern agricultural landscapes are in a constant flux of change and therefore it is necessary to look at the dynamics of the species as well.

Landscape ecology has been applied to many planning issues because of the relevance of the results to the planning process. Currently, in Holland, Belgium, Germany, and Denmark there is significant input from ecological work at the landscape level into the landscape design and management planning process to improve both the ecology and variety of the landscape and also the dispersal of species through that landscape. Schaller (1986) and Schaller and Haber (1988), in their work on the Bavarian geographical information system, also show how data structured at the landscape ecological level can be integrated into local plans. Primdahl (1985) and Agger and Brandt (1988) also demonstrate the interrelationships between landscape elements and land use and management in Denmark, and they emphasize the need for the coordination of nature conservation and agricultural management. The papers by Jongman (1984, 1990) on the Rhine catchment and on small streams in Holland demonstrate conclusively the ecological significance of the results of a consistent integrated approach and its links with the planning process. The core of the Dutch Nature Policy Plan (1990) is an ecological network consisting of core areas, buffer zones, and ecological corridors.

In terms of integration and coordination, the land classification procedure described by Bunce and Heal (1984) represents an important opportunity to synthesize data. The land classes themselves are an attempt to provide an overall summary of the environment at a local level which then holds correlations with other landscape factors. In the procedure progressive levels of ecological information are limited to define the characteristics of a range of landscapes.

CONCEPTS

The underlying concepts are important and have been much discussed within the field of landscape ecology. These can be ordered in a hierarchy from more or less general and holistic to more specific landscapes or population-oriented ones.

Sustainability

Sustainability is the capacity of the earth to maintain and support life and to persist as a system. The concept of sustainability is not only fundamental to the earth as a whole, but also to smaller systems within it. This parallels the approach of landscape ecology, in that it is essential to maintain ecosystems which are self-reproducing and not losing nutrients or species.

Sustainability implies that it is necessary to maintain a resource, whether it is wildlife, amenity, or ecological. The good husbandry concept of farming in the 19th century is in many respects reflected in some of the recent landscape ecological work on modern agroecosystems. An analysis of problems of sustainability has been carried out by IIASA (Clark and Munn, 1986) and brought to public awareness by the World Commission on Environment and Development (1987).

Hierarchy in Landscapes

Landscapes operate at different levels involving complexes of different elements. On the one hand, one can study a whole catchment such as the Rhine catchment (385,000 km^2) or, on the other hand, within that landscape one can examine structures such as a woodland and its surrounding land covers and their relationships. The Rhine catchment consists of mixtures of whole landscapes from Alpine to mountainous landscapes with large-scale forestry and mixed farming through to alluvial and lowland landscapes characterized by intensive dairy farming.

Gradients

A wider basic principle is that landscapes involve gradual changes and ecotones (Naiman and Decamps, 1990). It is recognized that many ecological elements do not show sharp boundaries between each other; rather, they grade gradually in time and space. The importance of edge effects has also been an integral feature of many studies with increases in diversity and structure. The stability and dynamics of such systems are based on physical parameters rather than on biological ones. This concept has been used in planning and nature conservation, but is not yet well supported by research.

Biodiversity

With the increased pressure on seminatural habitats there has been much concern shown about biodiversity. It is a basic concept in the management of landscapes and in planning. Policy objectives for natural parks and nature reserves are often formulated with the objective of maintaining an existing high biodiversity. Diversity is the outcome of historic processes and therefore refers to both time- and space-related processes (Pineda, 1990). Human activities can disturb or maintain high biodiversity, depending on the interaction of man with nature, especially by agricultural practices. Many natural and seminatural ecosystems, which once covered large areas, have been fragmented and their species are in danger.

Metapopulation

The very important landscape ecological concept concerning population dynamics in manmade landscapes is that of the metapopulation (Opdam, 1988). This represents the concept of the interrelationships between subpopulations in more or less isolated patches within a landscape and helps to understand the impact of progressive isolation of individual areas of vegetation and their associated animal populations in modern agricultural landscapes. Temporary extinction and recolonization are characteristic processes in metapopulations. In this respect, three aspects are important:

1. The dynamics of the subpopulations (extinction and immigration rate). If a patch is small and highly isolated, the extinction rate might exceed recolonization and a subpopulation becomes extinct.
2. The connectivity between patches. Important landscape variables in this respect are the absence of barriers and the presence of corridors.
3. The spatial and temporal variation in habitat quality. This is influenced by the absence or presence of disturbances in agricultural landscapes represented by land use practices.

DATA CAPTURE AND ANALYSIS

The above concepts require supporting data, which are usually acquired using objective methods, mainly because many studies involve planning agencies who need to have estimates of statistical reliability. Computer methods are widely applied, e.g., Geographical Information Systems, as these are very suitable for the display and analysis of data at the landscape level. Other supporting activities widely involved, are the development of scenarios of change, the design of landscapes and the integration of many different types of data applicable at the landscape level.

FUTURE DEVELOPMENTS

Landscape ecology is a developing science that presents many challenges, especially in the analysis of interactions between the elements that make up the landscape. The processes of intensification of agriculture in the lowlands and the contrasting decline in many upland areas, emphasise the imperative nature of applying landscape ecological principles in the development of appropriate maintenance policies for the countryside.

REFERENCES

Agger, P. and Brandt, J. (1988). Dynamics of small biotopes in Danish agricultural landscapes. *Landscape Ecol.* 1(4):227–240.
Andrewartha, H. G. and Birch, L. C. (1984). *The Ecological Web.* University of Chicago Press, Chicago, 506.
Asselin, A. (1988). Changes in grassland use: consequences on landscape patterns and spider distribution. *Connectivity in Landscape Ecology,* K.-F. Schreiber (Ed.), 85–88. Proc. 2nd Int. Semin. Int. Assoc. Landscape Ecol.
Baudry, J. (1984). Effects of landscape structures on biological communities: the case of hedgerow network landscapes, *Methodology in Landscape Ecological Research and Planning, Vol. I,* J. Brandt and P. Agger (Eds.), Roskilde Universit Centre, Denmark, 55–65.

Bunce, R. H. G. and Heal, O. W. (1984). Landscape evaluation and the impact of changing land-use on the rural environment: the problem and an approach. *Planning and Ecology, The Problem and An Approach*, R. D. Roberts and T. M. Roberts (Eds.), Chapman and Hall, London, 164–188.
Burel, F. (1988). Biological patterns and structural patterns in agricultural landscapes. *Connectivity in Landscape Ecology*, K.-F. Schreiber (Ed.), 107–110. Proc. 2nd Int. Semin. Int. Assoc. Landscape Ecol.
Burel, F. (1989). Landscape structure effects on Carabid beetles spatial patterns in western France. *Landscape Ecol.*, 2(4):215–226.
Burel, F. and Baudry, J. (1990). Structural dynamic of a hedgerow network in Brittany France. *Landscape Ecol.*, 4(4):197–210.
Clark, W. C. and Munn, R. E. (Eds.) (1986). *Sustainable Development of the Biosphere*, Cambridge University Press, Cambridge, 491.
Den Boer, P. (Ed.) (1971). *Dispersal and Dispersal Power of Carabid Beetles*, Miscellaneous Papers 8, Agricultural University, Wageningen, The Netherlands, 150.
Den Boer, P. (1977). *Dispersal Power and Survival. Carabids in a Cultivated Countryside*, Miscellaneous Papers 14, Agricultural University, Wageningen, The Netherlands, 190.
Forman, R. T. T. and Gordon, M. (1986). *Landscape Ecology*, John Wiley & Sons, New York, 619.
Harms, W. B. and Knaapen, J. P. (1988). Landscape planning and ecological infrastructure: The Randstad Study. *Connectivity in Landscape Ecology*, K.-F. Schreiber (Ed.), 163-167. Proc. 2nd Int. Semin. Int. Assoc. Landscape Ecol.
Harms, W. B. and Opdam, P. (1990). Woods as habitats for birds: application in landscape planning in The Netherlands. *Changing Landscapes, An Ecological Perspective*, I. S. Zonnerveld and R. T. T. Forman (Eds.), Springer-Verlag, New York, 73–97.
Harris, L. D. (1984). The fragmented forest. *Island Biogeography Theory and the Preservation of Biotic Diversity*. University of Chicago Press, Chicago, 211.
Jongman, R. H. G. (1984). The Rhine ecosystem, developments in planning and research. *Proc. 1st Int. Semin. Methodol. Landscape Ecol. Res. Plann.* Theme IV: Methodology of evaluation/synthesis of data in landscape ecology, P. Agger and J. Brandt (Eds.), 57–68.
Jongman, R. H. G. (1990). Conservation of brooks in small watersheds: a case for planning, *Landscape Urban Plann.*, 19:55–68.
Jongman, R. H. G., Ter Braak, C. J. and Van Tongeren, O. F. R. (1987). Data Analysis in Community and Landscape Ecology, PUDOC, Wageningen, The Netherlands, 299.
Mader, H.-J., Klüppel, R. and Overmeyer, H. (1986). Experimente zum Biotopverbundsystem-tierökologische Untersuchungen an einer Anpflanzung. Bundesforschungsanstalt fuer Naturschutz und Landschaftsökologie. *Heft*, 27, 136.
Miklos, L. (1989). The general ecological model of the Slovak Socialist Republic — methodology and contents. *Landscape Ecol.*, 3(1):43–51.
Ministry of Agriculture Nature Management and Fisheries (1990). Nature Policy Plan of The Netherlands. 103.
Neef, E. (1981). Stages in the development of landscape ecology. In: *Perspectives in Landscape Ecology*, S. P. Tjallingi and A. A. de Veer (Eds.), 19–27. Proc. Int. Congr. Organized by The Netherlands Soc. Landscape Ecol., PUDOC, Wageningen.
Naiman, R. J. and Decamps (Eds.) (1990). The ecology and management of aquatic-terrestrial ecotones. *Man and the Biosphere Series*, Vol. 4., Parthenon, 316.

Naveh, Z. and Lieberman, A. S. (1984). *Landscape Ecology. Theory and Application.* Springer-Verlag, New York.

Opdam, P. (1987). De metapopulatie: model van een populatie in een versnipperd landschap. *Landschap,* 4(4):289–306.

Opdam, P. (1988). Populations in fragmented landscape. *Connectivity in Landscape Ecology,* K.-F. Schreiber (Ed.), 75–79. Proc. 2nd Int. Semin. Int. Assoc. Landscape Ecol. Münster, 1987. Münstersche Geogr. Arb. nr 29.

Pineda, F. D. (1990). Conclusions of the international symposium on biological diversity, Madrid 1989. *J. Vegetation Sci.,* 1:711–712.

Primdahl, J. (1985). Agricultural, Wildlife and Landscape in Denmark. Recent trends in public debates, organizational influence and policy making. Inst. of Town and Country Planning, Royal Veterinary and Agricultural University, Copenhagen.

Risser, P. G., Karr, J. R. and Forman, T. T. (1983). *Landscape Ecology, Directions and Approaches.* Illinois Natural History Survey Spec. Publ. No. 2, 18.

Ruzicka, M. and Miklos, L. (1981). Methodology of ecological landscape evaluation for optimal development of land. *Perspectives in Landscape Ecology,* S. P. Tjallingi and A. A. de Veer (Eds.), 99–107. Proc. Int. Congr. Organized by the Netherlands Soc. Landscape Ecol., PUDOC, Wageningen.

Schaller, J. (1986). Ecological Balancing as a Tool for Environmental Impact Analysis. Rep. EEC-Workshop on the computerization of land use data — agricultural and environmental aspects. Federal Ministry of Food, Agriculture and Forestry, Bonn.

Schaller, J. and Haber, W. (1988). Ecological balancing of network structures and land use patterns for land-consolidation by using GIS-technology. *Connectivity in Landscape Ecology,* K.-F. Schreiber (Ed.), 181–187. Proc. 2nd Int. Semin. Int. Assoc. Landscape Ecol.

Schreiber, K.-F. (Ed.) (1988). *Connectivity in Landscape Ecology.* Proc. 2nd Int. Semin. Int. Assoc. Landscape Ecol. Münster, 1987. Münstersche Geogr. Arb. 29, 255.

Sukachev, V. N. and Dylis, N. V. (1964). *Fundamentals of Forest Biocoenology.* Oliver and Boyd, Edinburgh, 672.

Tansley, A. G. (1935). The use and abuse of vegetational concepts and terms. *Ecology,* 16:284–307.

Troll, C. (1939). Luftbildplan und ökologische Bodenforschung. *Z. Ges. Erd. Berl.,* 241–311.

Urban, D. L., O'Neill, R. V. and Shugart, H. H. (1987). Landscape ecology. A hierarchical perspective can help scientists understand spatial patterns. *Bioscience,* 37(2):119–127.

Van Dorp, D. and Opdam, P. (1987). Effects of patch size isolation and regional abundance on forest bird communities. *Landscape Ecol.,* 1(1):59–73.

Verboom, B. and Van Apeldoorn, R. (1990). Effects of habitat fragmentation on the red squirrel *Sciurus vulgaris* L. *Landscape Ecol.,* 4(2/3):171–176.

Vink, A. P. A. (1983). *Landscape Ecology and Land Use.* Longman, London, 264.

Webb, N. (1986). *Heathlands. A Natural History of Britain's Lowland Heaths.* Collins, London, 223.

World Commission on Environment and Development (1987). *Our Common Future.* Oxford University Press, Oxford, 400.

Zonneveld, I. S. (1972). Landevaluation and land(scape)science. ITC Textbook of Photo-Interpretation Vol. 7. ITC Enschede, 106.

Zonneveld, I. S. and Forman, R. T. T. (Eds.) (1990). *Changing Landscapes, An Ecological Perspective.* Springer-Verlag, New York, 286.

2 The Ecological Significance of Linear Features in Agricultural Landscapes in Britain

R. G. H. BUNCE
C. J. HALLAM

Abstract: *A series of ecological surveys in Britain have examined linear and open landscape habitats in the rural countryside, in order to determine their ecological significance. Botanical data are presented for verges, hedges, and streamsides recorded in representative sample areas in British landscapes in 1978. The analysis of these data showed that linear features contained many species that were absent in the open fields of the lowlands, whereas in the uplands species are distributed more evenly through the landscape. In 1988 further linear features were sampled, e.g., walls and tracks, and results showed that in the lowlands the range of linear features present added further emphasis to the initial conclusions regarding their importance. These analyses formed the basis for the design of the methodology for a further survey of the British countryside in 1990, the results of which are currently being analyzed. It is concluded that linear features are important reservoirs for the conservation of vegetation and are also potential sources of propagules for recolonization of agricultural land where the majority of plant species have been lost.*

INTRODUCTION

Over the last 15 years, a series of broad-scale ecological surveys have been carried out throughout Britain by the Institute of Terrestrial Ecology. These surveys were designed to describe and characterize environmental strata termed land classes from samples of vegetation. These strata were derived from mapped environmental data using multivariate analysis. A subsample of 1-km squares from the strata were then surveyed in the field. In the initial survey carried out in Cumbria (Bunce and Smith, 1978), 16 random quadrats of 200 m^2 were placed within representative 1-km squares. During the field survey, it was observed that quadrats placed in open country missed much of the variation that was apparent in linear features, such as streamsides and roadside verges. Therefore, in 1978, when the land classification method was extended to cover the whole of Great Britain, data from three linear features were recorded.

Since the analysis of these data, a wider recognition of the significance of linear features has developed, with the incorporation of concepts from landscape ecology relating to connectivity and corridors. This is reflected in a wide range of literature now available on the subject, e.g., Schreiber (1988) and Bunce and Howard (1990).

In 1988, the study of linear features was extended by recording species from all linear features within a sample of 1-km squares. The results from the surveys in 1978 and 1988 were used to indicate the broad ecological composition of linear features in Britain.

METHODOLOGY

In 1978, the first survey of Britain was designed to provide quantitative data on the distribution of vegetation and habitats in Britain. A field survey of random 1-km squares drawn from 32 environmental classes (Bunce and Smith, 1978; Bunce and Heal, 1984) recorded vegetation data from five 200-m^2 quadrats placed at random in the 1-km square, plus two 10-m^2 quadrats placed along of streams, verges, and hedges. In 1988 complete species lists were recorded from linear and areal habitats within a subsample of 1-km squares.

The quadrat data from 1978 have been analyzed using multivariate statistics, mainly TWINSPAN (Hill, 1979a) and DECORANA (Hill, 1979b), to produce species groups and vegetation types to assess the botanical diversity and composition of the linear features. All the data are presented by land classes, which are the environmental strata described by Bunce et al. (1981). The classes are ordered according to the dominant land cover, so that those with lowland vegetation are on the left and the most upland on the right. The first group of classes is dominated by cereal crops, the second by lowland grasslands, the third by the mixed vegetation of the marginal uplands, and the fourth by open moorlands in the hills.

RESULTS

Figure 1 shows the distribution of vegetation types, as identified by TWINSPAN, in the different land classes. The number of vegetation types in the hedgerows is variable in the lowlands, being related to local land management and environmental factors, but with no obvious relationship to the principal environmental gradient. The length of hedgerow is related to this gradient, hedges being most frequent in the grassland areas in the west and less prevalent in the north and uplands.

Verges are less variable than the hedges in the lowlands and in the uplands are relatively homogenous. This is because the lowland verges are often relicts of less intensively managed grasslands which have now disappeared from the surrounding landscapes, whereas in the uplands they are intrinsically less variable.

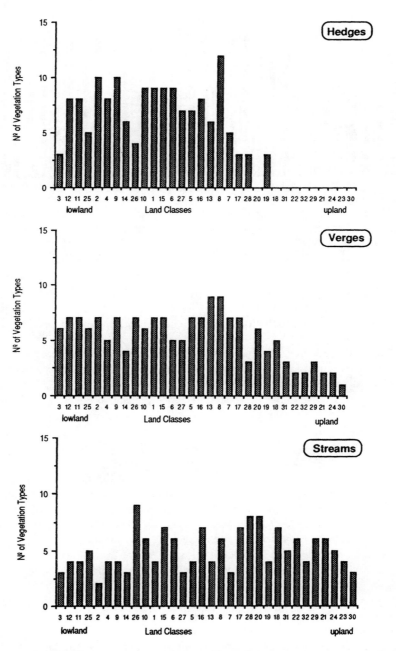

Figure 1. The number of vegetation types recorded throughout Britain in 1978, using quadrats placed along linear features. Two quadrats were placed at random along the three linear features in each of 8 1-km squares. These squares were drawn at random from 32 land classes. The vegetation types were produced by analysis of species data using TWINSPAN.

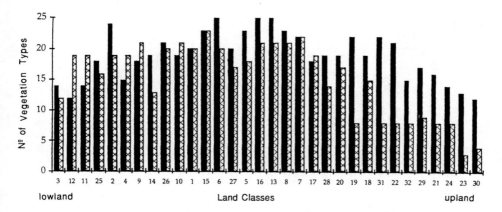

Figure 2. The occurrence of vegetation types in open habitats and linear features throughout Great Britain in 1978. The data from 2 10-m² plots placed on each of the 3 linear features are compared with data from 5 200-m² quadrats, placed at random in the same 1-km squares. ■ Open habitats; ⊠ linear features.

The streamsides are relatively uniform in the lowlands because of the effects of eutrophication and stream bank management. In the uplands, diversity of streamsides is greater due to localized enrichment and, in some cases, protection from grazing.

Thus, the three linear features show distinct ecological patterns; hedges and verges contribute most diversity in the lowlands, while streams are a more significant source of variation in the uplands.

In Figure 2 the number of vegetation types present in all three linear habits is compared with the number found in the open countryside. The latter show least diversity in the uplands and in arable-dominated lowlands. Diversity is greatest in the marginal uplands land where both upland and lowland habitats are present. In contrast, linear features are more diverse in the lowlands and are relatively uniform in the upland landscapes of northern Britain. There is a marked contrast between the distribution of vegetation types in the lowlands and the uplands, with the former having most variation restricted to linear features whereas in the latter there is less distinction. Within this general pattern there is much local variation, e.g., the seminatural vegetation on chalk and limestone is diverse in areas where it has not been replaced by intensive agriculture. Such patterns are the result of historic changes in land use and would have been very different only 50 years ago.

Figure 3 illustrates the diversity of linear features, in terms of numbers of plant species. In arable-dominated areas, hedges have frequently been removed; where they are still present, the herb layer is often impoverished by herbicide spraying and eutrophication. The number of plant species is higher in the marginal uplands where the hedges are associated with pasture and so have been less disturbed.

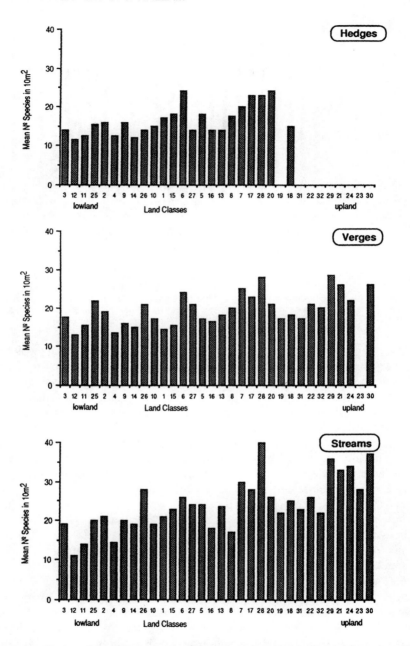

Figure 3. The number of species recorded from linear features throughout Britain in 1978. Two randomly located 10-m² plots were placed along each of the 3 linear features in 8 1-km squares. These squares were drawn at random from the 32 land classes.

LANDSCAPE ECOLOGY and AGROECOSYSTEMS

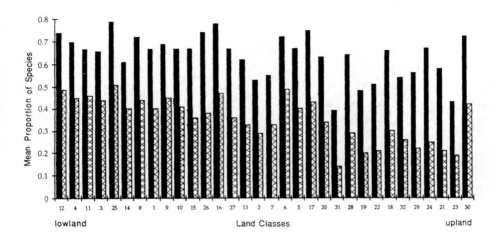

Figure 4. The proportion of plant species recorded in linear features in Britain in 1978, and the proportion found in linear plots only. The data from 2 10-m² plots along each of 3 linear features are compared with data from 5 200-m² quadrats. These quadrats were placed at random in the same 1-km squares. ■ All species found in linear features; ▨ species found early in linear features.

The verges show a slight trend towards greater species numbers in the upland classes — in contrast to the diversity of vegetation types, showing that more than one measure of diversity is required to express the variation present.

The streams show a greater degree of variation overall, but also have more species in the uplands. This is due to localized flushing by streamwater and the more variable physical environment.

In Britain as a whole, the streams have more species than the verges, with hedges having the fewest. Previously, most emphasis has been given to hedgerows, but these data show that other linear features can be more significant.

Each of the linear features investigated contains distinctive species not found in the open countryside. This is especially so in the case of streamsides, which provide a habitat for wetland species, e.g., *Caltha palustris, Iris pseudacorus,* and *Glyceria fluitans*. Hedges contain many woodland species such as *Digitalis purpurea, Tamus communis,* and *Silene dioica,* while verges contain species of mesotrophic grasslands such as *Filipendula ulmaria, Geranium pratense,* and *Galium verum*. Thus linear features not only contain the species present in the open landscape, e.g., *Lolium perenne* and *Agropyron repens,* but, in addition, have their own complement of distinctive species. Although only six 10-m² plots were recorded in the linear features as opposed to the five 200-m² in the areal habitats, Figure 4 shows that the plots from linear features contain a high proportion of all the species recorded in the 1-km squares. In addition, many of these species were not found at all in the fields. In the uplands, plant species are more evenly distributed through the landscape than in the lowlands, so the linear features are less distinctive.

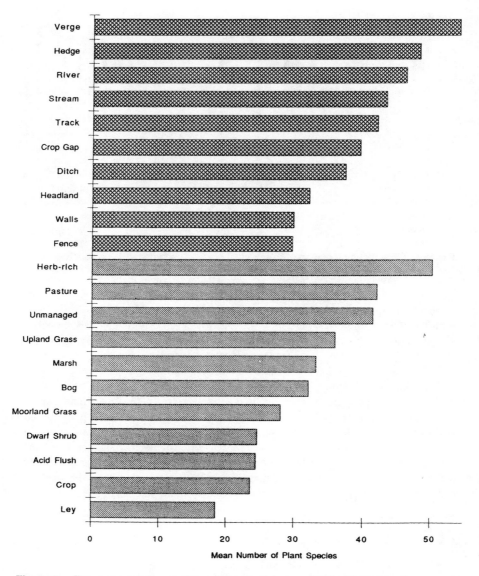

Figure 5. The mean number of higher plant species recorded in different habitats, from 64 1-km squares, throughout Britain, in 1988.

Figure 5 shows the mean number of species found in each habitat in 1-km squares sampled throughout Britain. In general, the linear features contain more species per unit area than habitats in the open countryside. The majority of the species-rich linear features are in the lowlands where the open countryside contains very few species. Even in gaps between the crop and the field boundaries

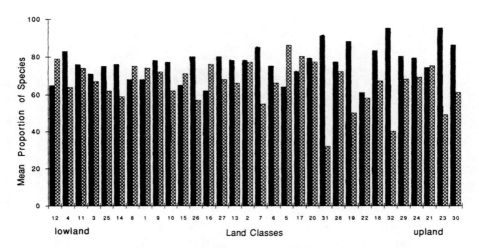

Figure 6. The mean proportion of higher plant species, recorded in 1-km squares, which occurred in linear features, and open habitats. From data for 64 1-km squares, throughout Britain, in 1988. ■ Open habitats; ▩ linear features.

there are unsprayed areas where uncommon arable weeds, which have been eliminated from the crop, persist, e.g., *Thlaspi arvense* and *Anchusa arvensis*. These fragmented patches will contribute the majority of species available for recolonization if extensification takes place, emphasizing that the management of linear features in lowland landscapes is critical for the maintenance of diversity.

This point is confirmed by Figure 6, which shows that, in the lowlands, there are as many species present in the linear habitats as in the open landscapes, although the former occupy less than 5% of the land area. In contrast, in upland areas the majority of species are widely dispersed throughout the landscape, in part because most of the open habitats are less intensively managed, and in part because the linear habitats are less common. In general, the lowland habitats are intrinsically richer in species than the uplands, but intensive management has meant that many plant species are restricted to small patches of the landscape.

CONCLUSIONS

The results presented above indicate the botanical significance of linear features in the British countryside. The importance of these features is further emphasized by the paucity of undisturbed seminatural vegetation in intensively farmed lowland landscapes.

Linear features are not only important for flora but also for fauna. Some species such as the phytophagous insects are directly dependent upon plants; others depend upon them for shelter and movement. Examples of these interrelationships have been presented by Bunce and Howard (1990) and Schreiber (1988).

Linear features are also important for the role they play in linking elements of the landscape, especially where seminatural habitats are small and fragmented. Dispersal along linear features has been widely observed previously, e.g., the spread of *Senecio squalidus* along railway lines and *Chamaenerion angustifolium* along verges to colonize marginal derelict areas. More recently, oilseed rape has been seen spreading along roadside verges. This potential for dispersal, provided by roadside verges, may allow movement of species following climate change; for example, the major drought of 1990 in southern England killed much roadside vegetation, producing bare areas which became available for colonization.

The role of many linear features, as a refuge for species which have not been able to survive in the fields, is important in terms of the response of vegetation to changes in agricultural practices, such as set-aside. For example, a decline in management of grasslands could lead to the expansion of species from field boundaries, as seen in the derelict areas in the Pyrenees. The mobility of different species and their ability to colonize existing vegetation will determine which are successful.

Linear features still retain much botanical capital that can be utilized to replace species lost from open landscapes. They need, therefore, to be incorporated into the development of landscape design, for maintaining and enhancing ecological diversity. While it is recognized that designated conservation areas for specific habitats are essential, the above work shows that even intensively farmed landscapes still contain a surprising range of species that offer opportunities for conservation.

REFERENCES

Bunce, R. G. H. and Smith, R. S. (1978). An Ecological Survey of Cumbria. Working paper No. 4. Cumbria County Council and Lake District Special Planning Board.

Bunce, R. G. H. and Heal, O. W. (1984). Landscape evaluation and the impact of changing land use on the rural environment: the problem and an approach. *Planning and Ecology*, R. D. Roberts and T. M. Roberts (Eds.).

Bunce, R. G. H. and Howard, D. C. (1990). *Species Dispersal in Agricultural Habitats*. Bellhaven, London.

Bunce, R. G. H., Barr, C. J. and Whittaker, H. A. (1981). The land classes of Great Britain: preliminary descriptions for users of the Merlewood method of land classification (Merlewood Research & Development Paper, No. 86), Grange-over-Sands: Institute of Terrestrial Ecology.

Hill, M. O. (1979a). DECORANA — a Fortran Program for Detrended Correspondence Analysis and Reciprocal Averaging. Cornell University, Ithaca, NY.

Hill, M. O. (1979b). TWINSPAN — a Fortran Program for Arranging Multivariate Data in an Ordered Two-Way Table by Classification of Individuals and Attributes. Cornell University, Ithaca, NY.

Schreiber, K. F. (Ed.) (1988). *Connectivity in Landscape Ecology*. Proc. 2nd Int. Semin. Int. Assoc. Landscape Ecol. Munster, Schonigh, Paderborm.

3 Landscape Dynamics and Farming Systems: Problems of Relating Patterns and Predicting Ecological Changes

JACQUES BAUDRY

Abstract: *The results of two studies are presented. The first concerns land use dynamics in western France during the last 100 years; the second examines the relationships between farming systems, the physical environment, and landscape patterns in a municipality. Analysis shows that predictions of changes in agricultural landscape patterns, and thus their ecological consequences, are due to scale-dependent phenomena as well as to interactions among various levels of organization of the economic, social, and ecological systems. In the municipality studied here, farmer's decisions explain landscape patterns to a greater extent than physical constraints, although only 50% of the land is arable. In the discussion it is suggested that fractal geometry may be useful to make predictions across spatial levels as well as across time scales.*

INTRODUCTION

In agricultural landscapes, land use (types of crops, management practices) and landscape structure (spatial arrangement of landscape elements) are important factors in the control of species distribution and ecological processes (Zonneveld and Forman, 1989).

Both land use and landscape structure are under the control of farmers and planners who make decisions within a set of physical and socioeconomic constraints. Potential techniques often rely on the assessment of soil capability for agriculture (Flaherty and Smit, 1982), ignoring the economic environment. But studies of land use dynamics (INRA-ENSSAA, 1977; Bazin et al., 1983; Iverson, 1988) reveal the importance of human decisions in the formation of landscape patterns. Phipps et al. (1986) showed that land use changes around Ottawa between 1955 and 1978 are related to both soil characteristics and distance to urban areas. Generally, changes in the economic context lead to changes in the perception of physical constraints by farmers (Baudry, 1991). Golley and Ryszkowski (1988) stress the importance of socioeconomic factors as control variables of farming systems and of the ecology of rural areas. At a given time, broad-scale (regional) socioeconomic variables are similar for different landscapes, but they differ on a fine scale (farm level) where variables such as the farmer's age and farm size can explain differences between landscape patterns.

In this chapter, a brief overview is presented of land use changes in western France in the last century, followed by a case study of interactions between farmers and the landscape. In conclusion, some conceptual and methodological problems posed to ecologists regarding such changes are examined.

It has been proposed that landscapes should be considered as fractal objects (Loehle, 1983; Mandelbrot, 1984; Milne, 1988), whose find grain structure may interlock with a coarser grain structure over a certain range of scales. Within a fractal domain, it is possible to transfer observations across spatial scales (and possibly across time scales if rates of changes are fractal) (Milne, 1991; Sugihara and May, 1990). As fractal geometry may be a possible framework to deal with the problem of scale, fractal patterns are examined.

LAND USE CHANGES IN WESTERN FRANCE SINCE 1882

The changes are assessed within census units which are administrative. The smallest unit is the municipality, then comes the "département" which is about 600,000 ha. The départements (e.g., Manche, Calvados, and Morbihan) are grouped into administrative regions (e.g., Brittany and Lower Normandy).

Land use has changed dramatically in western France during the last hundred years; for example, in the Manche département grassland covered 20% of the total surface in 1882, almost 75% in 1965, and then back to 54% in 1985. Such a shift in land cover probably had many ecological implications that we are not able to assess. As more changes are forecast, due to changes in the European Economic Community (EEC) agricultural policy, it is important to examine the patterns of these changes to see if predictions can be made and to propose further ecological analysis.

Material and Methods

Annual censuses were supplied by the Ministry of Agriculture for 1882, 1911, 1920, 1939, 1955, 1965, 1975, and 1985 for two regions and their départements: Brittany (Côtes d'Armor, Finistère, Ille et Vilaine, and Morbihan) and Lower Normandy (Calvados, Manche, and Orne). Four types of land use were considered in the analysis: plowland, grassland, moorland, and woodland (a distinction is made between arable land — land that can be plowed, and plowland — land that is actually ploughed).

Frequencies were extracted from each département to give the same load in the factorial correspondence analysis that yields gradients of land use within which trajectories of changes can be described (Benzécri 1973; Burel and Baudry 1990). Brittany and Lower Normandy, as well as the whole study area, were mapped as supplementary elements in the factorial space. This method is useful

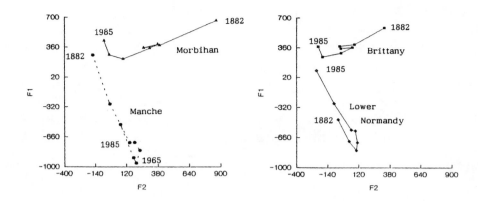

Figure 1. Examples of trajectories of land use changes in western France in two départements (Morbihan and Manche) and two regions (Brittany and Lower Normandy). F1 and F2 are the first and second factors of a correspondence analysis. Along F1, grassland cover decreases from negative to positive values. Along F2, moorland decreases from positive to negative values.

to represent various hierarchical levels in the same reference frame (Burel and Baudry, 1990). The rates of the changes were computed at various spatial scales during the period.

Results: Trajectories and Rates of Change

The first two factors of the ordination explain 94% of the total variance. The eigen value on the first factor was 0.24, which indicates a rather smooth land use gradient. The first gradient indicates the importance of the plowland as opposed to grassland (plowland up to 70% in the positive end of axis one, grassland up to 80% in the negative part). The second gradient indicates the importance of moorland (up to 40% in the positive part of the axis). The important changes noted above are clear on the factorial map (Figure 1). Brittany and Normandy exhibit different patterns of change, with a change of tendency for Normandy in the 1960s. The trajectories are irregular, but get smoother at higher levels in the spatial hierarchy (Figure 1) because rates of change decrease as the scale gets coarser. The départements (fine scale) seem to change faster than the regions (coarse scale) (Figure 2). The rates of change are therefore scale dependent.

This phenomenon of scale dependence also occurs when the rates of the changes are considered over a range of the time scale (Baudry, 1991). This result is important, if not surprising, since scale dependence is frequent in ecological systems (Wiens, 1989). The results are also consistent with the hierarchy theory (Allen and Starr, 1982; O'Neill, 1989) which predicts that systems that involve several components behave more slowly than lower levels.

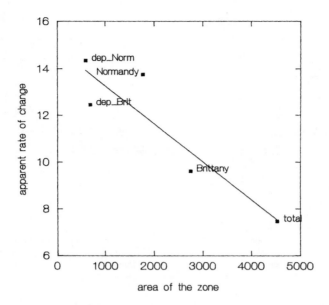

Figure 2. Rates of land use changes at different spatial levels (dep_ Brit = départements from Britanny, dep_ Norm = départements from Normandy).

FACTORS OF CONTROL AT THE LANDSCAPE LEVEL

The goal of the research was to understand the relative importance of socio-economic as opposed to ecological variables in the dynamics of landscape change.

Study Site

The research was carried out in Sainte-Marguerite de Viette, a 800-ha municipality in the Pays d'Auge, Central Normandy, in western France. Changes in landscape are seen, such as the extension of annual crops at the expense of grassland. Poor management of grassland is also present in some meadows, though truly abandoned land is uncommon. There are 41 farmers who use land in the municipality.

Material and Methods

The municipality was investigated by surveys conducted among all users of any agricultural field (i.e., 90% of the municipality area) and field surveys, including an assessment of land agricultural capacity carried out by the group of farmers themselves. Grassland is the dominant land cover (79%) as cattle raising exists on all the farms, but only one farmer out of three raises dairy cows. Data were recorded on a grid of 0.25 ha (see list of variables, Table 1)

Table 1 List of Variables Used in the Analysis

Description of the physical environment and landscape patterns in 16-ha windows. In the data matrix the variables are coded in 3 classes: (1) less than 25% of the 0.25-ha cells, (2) 25–50%, or (3) more than 75%

Variables for the physical environment

Land capability
    Arable land class 1 (prime land)
    Arable land class 2 (average land)
    Arable land class 3
    Nonarable land class 1
    Nonarable land class 2
    Nonarable land class 3
    Nonarable land class 4

Topography (contours are 5 m apart)
    Flat land (0 or 1 contour)
    Gently rolling (2 contours)
    Smooth slope (3 contours)
    Slope (4 contours or more)

Variables for landscape patterns

Land cover
    Grassland
    Plowland
    Ungrazed patches
    Woodlot

Hedgerows
    Number of hedgerows in the 0.25-ha cells
    Number of connections among hedgerows
    Number of tracks bordered by two hedgerows

Description of farming types: each 0.25-ha cell is characterized by the following classes of variables:

    Farmer less than 35 years old
    Farmer more than 35 and less than 55 years old
    Farmer more than 55 years old
    Farm less than 10 ha
    Farm 10 to 20 ha
    Farm 20 to 50 ha
    Farm more than 50 ha
    Less than 10% of the farm plowed
    10 to 50% of the farm plowed
    More than 50% of the farm plowed
    Dairy cows less than 4% of the herd
    Dairy cows between 4 and 40% of the herd
    Dairy cows more than 40% of the herd
    Herd less than 10 cattle units
    Herd 10 to 40 cattle units
    Herd more than 40 cattle units

The 16-ha windows are characterized by the number of 0.25-ha cells in each class.

to enable study of the relationships in space between the variables. Through the use of moving windows (Milne, 1991) the landscape was examined at different scales (spatial resolution) by using windows of 1, 4, and 16 ha. A routine, using DBase IV, was developed by Denis (1991) to aggregate cells in the database in order to perform analyses at various scales. In this routine all the fine-scale information is saved when recaling. If, in a 16-ha window, plowland is only 25% of the area, the window is characterized by 25% plowed land and 75% grassland. This contrasts with another routine (e.g., Turner, 1990) where only the dominant land cover is retained through aggregations. In this paper, landscape pattern only is considered, so results are only presented from study of the 16-ha windows (575 windows, from 16 grids).

Data analysis includes a correspondence analysis of the variables related to landscape characteristics (Table 1). A cluster analysis was carried out to describe the types of landscape patterns at the 16-ha window scale. A similar analysis was carried out on the physical variables, to define the types of physical environment.

The Special Case of Farming Systems

The definition of a landscape pattern in sample squares of any size is no problem because landscape is continuous in space. This is not the case for farming systems because farms are of different sizes (from 5 to 120 ha, in this case) and discontinuous in space. Generally, a farming system (i.e., a farm with capital, machines, and objectives of production) does not match a landscape pattern. It is therefore difficult to link various standpoints on farming activity (Hubert and Girault, 1988). These patterns can only be correlated with the mixture of farming systems operating in the same area. Studies at the field scale may also explain landscape patterns (Baudry, 1989). In this paper, only very general structural variables are used to describe the farming systems (farmer's age, farm size, type of cattle, and importance of plowing; see Table 1). The rationale of the farmer's choices was not examined nor was the functioning of the various systems; this would need studies at the farm level on a fine time scale.

Farming systems have a hierarchical structure (field, farm, and a set of contiguous farms). Within the framework of hierarchy theory (Allen and Starr, 1982) structural variables that are constraints at the farm level can be seen as constant for the study of farm functioning. At the municipality level, and at the time scale of land use changes, they are dynamic and can explain landscape modification.

To describe the farming constraint in a landscape, a first option might have been to establish types of farming systems from multivariate analysis of data at the farm level. In this option each farm represents one individual in the analysis. However, it was decided to give each farm a weight proportional to its size in the definition of the farming systems. Each 0.25-ha cell is characterized by the variables (Table 1) of the farm it is included in. Farming types were obtained from the analysis of these data.

For farming variables, their co-occurrence at the 16-ha scale may make little sense; the real world is the value of variables within a farm, the combinations of these variables are found with other combinations within the 16-ha windows. Correspondence analysis was used on the farming variables in the 2492 0.25-ha cells and the 575 16-ha windows were put in as supplementary elements. The gradients therefore reflected real systems and their co-occurrence in space. Farming types are not actual systems, they are only combinations of different farming systems in a window.

The relationships between the landscape clusters and the characteristics of the farming systems operating within each window were then studied. ADDAD routines (Lebeaux, 1985) were used for multivariate analysis and SYSTAT/SYGRAPH routines (Wilkinson, 1988) for the study of relationships and mapping. In order to assess the relationships between these classifications, mutual information (or joint information) between them was computed (Phipps, 1981).

If L represents the landscape types, P the physical environment, and F the farming types, then **H(L)** is the diversity of landscape types calculated by Shannon's formula, **H(P)** is the diversity of physical environment types, and **H(F)** is the diversity of farming types. **T(L/F) = H(L) + H(F) − H(L,F)** is the mutual information between the landscape types and the farming types, i.e., the amount of information on the landscape types, knowing the type of farming in the same window. Kullback's test (Kullback, 1959) allows testing to see whether T is statistically significant or not. The test is similar to a chi^2 test, with no constraint on distributions. Kullback criterion K = 2 N T follow a chi^2 law at (c − 1)(r − 1) degrees of freedom, where c = number of colums and r = number of rows.

R(L/F) = T(L/F)/H(L) is a measure of the landscape diversity explained by the differences between farming types (Phipps, 1981). A high R indicates that farming types are a good predictor of landscape differentiation. **R(L/P) = T(L/P)/H(L)** gives the same information regarding the effects of the physical environment. R is a measure of "landscape organization", i.e., of our ability to make predictions on landscape patterns from other sources of information (Phipps, 1981; Phipps et al., 1986).

As it is a reasonable hypothesis that a single farming system will shape the landscape differently within different resources (types of physical environment), the same computation of R(L/F) was carried out within each type of physical environment.

Results: Physical Environment, Landscape Types and Types of Farming

The Physical Environment

The cluster analysis yielded three types of physical environment, five types of landscapes patterns, and four types of farming, which reflect the characteristics of the farms within each 16-ha window. The three types of physical environments differ according to the amount of arable land and/or of prime quality land. The

Table 2 The Main Characteristics of the Three Types of Physical Environment Derived From Cluster Analysis

Type	Number of cells	% Arable land	% Prime land	% Avg land	% Flat land
1	140 (24%)	20	11	08	58
2	248 (43%)	46	37	08	84
3	187 (33%)	72	05	63	84
Avg		48	20	26	77

Number of cells is the number of 0.25 ha cells and % of the study area; % avg. land is arable land of average quality. Avg gives the average characteristics of the study area.

Table 3 The Main Characteristics of the Five Types of Landscape Pattern Yield by Cluster Analysis

Type	Number of cells	Grassland (%)	Ungrazed (%)	Number of hedgerows	Number of connections
1	420 (73%)	83	04	56	22
2	105 (18%)	51	04	49	19
3	36 (6%)	71	02	88	42
4	10 (2%)	20	00	45	10
5	4 (0.5%)	58	00	2	0
Avg		75	03	56	35

Number of hedgerows is the sum of the number of hedgerows found in each 0.25 ha cell, it is an estimate of hedgerow density. Avg gives the average characteristics of the study area.

latter appears not to be concentrated in windows with the highest percentage of arable land (Table 2). Type 1 environments have major constraints.

The Landscape Patterns

Among the five types of landscape patterns, Type 1 covers almost 3/4 of the area and shows the average characteristics of the muncipality (Table 3); a high percentage of grassland and numerous hedgerows. Type 2 has only 50% of grassland. Type 3 has grassland and a dense hedgerow network, in terms of hedgerows and the connections among them. Type 4 and 5 are very infrequent: Type 4 is plowland and restricted to a small part of the municipality. Type 5 has almost no hedgerows; it is found in an area shared by one farmer raising horses on large meadows and another oriented toward cash crops.

The Types of Farming

Four main types of farming were recognized (Table 4); each occupies between 20 and 40% of the municipality:

- Type 1 is mostly beef production; the farmers are older than the average, with no or little plowland. Most of these farmers do not use fertilizer.
- Type 2 also have no or little plowland but the farmers are younger, with larger herds and with dairy cows. (These first two types are the archetypes of the agricultural systems of the Pays d'Auge.)

Table 4 The Main Characteristics of the Four Types of Farming Systems

Type	Number of cells	Grassland (%)	Plowed land (%)	Arable land (%)	Young farmers (%)	Middle-aged farmers (%)	Old farmers (%)	Farms 20–50 ha (%)	Farms >50 ha (%)	More than 40 cows (%)	With 10–40 cattle unit (%)	With more than 40 cattle units (%)
1	217 (38%)	80	10	47	22	14	45	50	13	06	45	22
2	127 (22%)	84	06	48	12	40	28	70	11	26	26	53
3	112 (19%)	58	33	55	56	11	14	25	50	06	55	25
4	114 (21%)	75	17	47	23	53	06	47	36	36	06	60
Avg		75	15	48	27	27	27	49	24	17	35	37

Avg gives the average characteristics of the study area.

- Type 3 areas are used by large farms with 30% of the land plowed; the farmers are young and have medium-sized herds with few or no dairy cows. These are the most intensive farms according to local standards.
- Type 4 represents another type of intensification: where dairy cows are fed with maize as silage, farms of of average size, and farmers are middle-aged, as in Type 2, but which, however, is more traditional. Middle-aged farmers took the opportunity to increase milk production before quotas were put on that production, whereas young farmers could not.

It is worth noting that these four types do not differ in the amount of arable land (roughly, 50% in each). Land quality is not a major factor determining the farming systems. This is also reflected in the fact that, in this muncipality, farming types are spatially distributed almost independently of the physical environment (R[F/P]) = 3.5%, significant, but very low). This means that currently farmers do not yet have a strategy of discrimination between the land capabilities of fields, as can be seen in the diversity of production within farms (e.g., cattle raising and cereal crops).

The Relationships

Table 5 gives the relationships between the various classifications. The study of the relationships between landscape patterns, the physical environment, and farming types reveals that differences among farming types explain twice as much landscape pattern diversity as differences in the physical environment: R(L/F) = 22.1% (p <0.001) and R(L/P) = 11.1% (p <0.001). This means that farmers' decisions are more important here than land capability for landscape patterns.

Farming Type 3 ("young cereal producer") is the only one in landscape 4 (plowed land dominant), and is absent in landscape 3 (grassland, dense hedgerow network). Farming Types 1 and 2 ("traditional") are almost restricted to landscapes 2 and 3 (grassland dominant, many hedgerows). Farming type 4 ("modern dairy") is mostly present in grassland-dominated areas. Land abandonment is not related to any farming type.

R(L/F) increases when computed within the different types of physical environment. It is 22.6% in Type 1 (highest constraints), 27.7% in Type 2, and 26.7% in Type 3. Farmers who grow crops plow most of the arable land. When the soil is not arable, they differ from the others by having larger meadows and thus fewer hedgerows.

Figure 3 represents the maps of the three classifications.

FRACTAL PATTERNS

These were examined at two levels: Lower Normandy for grassland and major landscape variables. The fractal dimension was estimated by the grid method

Table 5 The Relationship Between the Different Classifications On 16-ha Windows

Relationships Between the Physical Environment and Landscape Patterns

Type of landscape pattern	Type of physical environment			
	1	2	3	Total
1	124	194	102	420
2	12	39	54	105
3	4	1	31	36
4	0	10	0	10
5	0	4	0	4
Total	140	248	187	575

Relationships Between the Type of Farming and the Physical Environment

Type of Farming	Type of physical environment			
	1	2	3	Total
1	75	57	85	217
2	22	60	45	127
3	20	57	37	112
4	23	74	22	119
Total	140	248	187	575

Relationships Between the Landscape Patterns and the Type of Farming

Type of farming	Type of landscape pattern					
	1	2	3	4	5	Total
1	180	19	18	0	0	217
2	110	0	17	0	0	127
3	45	56	1	10	0	112
4	85	30	0	0	4	119
Total	420	105	36	10	4	575

The tables give the number of windows for each case.

(Voss, 1988), counting the number of windows where a pattern is present at the different scales. With moving windows, several grids exist for each scale (except the finest), and all were used in the study.

In lower Normandy, grassland distribution in space (Figure 4) is fractal. The fractal dimension of the pattern is different according to scale, because the variables that control the importance of grassland at the regional level differ from those at the microregional or landscape level. At the regional level climate and bedrock are the major constraints, and at the landscape level (i.e., within a given climate and bedrock) characteristics of the farming systems are discrim-

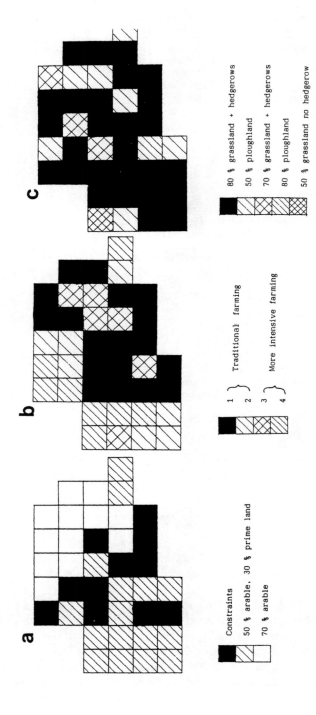

Figure 3. Maps of the (a) physical environment, (b) farming types, and (c) landscape. See tables for definitions of the types.

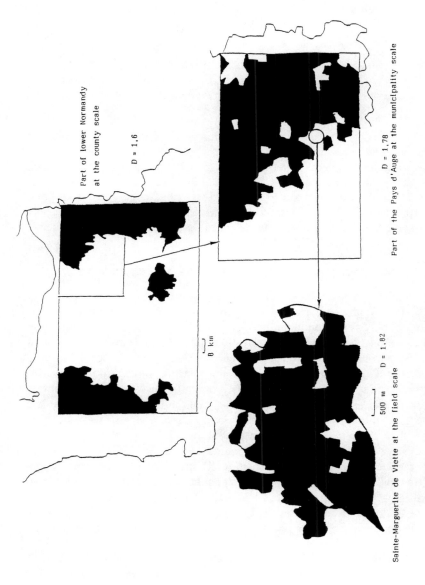

Figure 4. Fractal distribution of grassland at different spatial levels in Lower Normandy. Maps show units (counties, municipalities, and fields) covered by more than 80% of grassland. D is the fractal dimension of grassland at each level.

Table 6 Fractal Dimension (D) of Various Types of Spatial Patterns in Sainte-Margueritte-de-Viette

Type of pattern	D	Constant	r^2
Hedgerows	1.80	−0.871	0.99
Ungrazed patches	0.82	0.81	0.97
Arable land	1.72	−0.77	0.99
Plowland	1.27	0.22	0.99
Farmers <35 years	1.48	−0.34	0.99
Farmers >55 years	1.45	0.24	0.99
Farms >50 ha	1.41	−0.15	0.99

Computation of D: $\log(Ns) = D \log(S) +$ constant where S is the scale (number of windows on the side of the grid) Ns is the number of windows where the pattern is present.

inant. The high D at the landscape and microregional levels indicates that the grassland is in large, connected patches.

In the study area, several patterns have a fractal structure. They are natural patches (arable land), land use (plowland), and even distribution of social variables such as the farmers' age (Table 6). This result means that fine grain patterns (at the 0.25-ha scale) are similar to coarse grain ones (at the 16-ha scale). Figure 5 gives an example of the fractal structure of ungrazed land.

DISCUSSION: CHANGING LANDSCAPES AND FARMING SYSTEMS

Most predictions of land use change are made according to land capability. Farmers' choices, however, may be as important, but they are difficult to predict. They are not only dependent upon market conditions (which are uniform in the present case), but also upon farm size or soil quality. Farmers also make choices of land use according to personal preferences. The jagged shape of land use trajectories at a fine scale may be due to the fact that farmers do not change their systems simultaneously and that they proceed by successive trials before a clear trend is chosen. At a coarser time scale, trials and errors are smoothed and only the general trend is evident. Furthermore, there are lags between changes of national or international economic conditions and farmers' reactions. For example, since the inauguration of milk quotas within the EEC, in 1984, there are still farmers giving up milk production. **If these hypotheses prove to be generally true, detailed studies of short-term land use changes are of no use for predicting long-term change.** Changes in the price of farm products, land and labor costs, farm population, and employment opportunities in other economic sectors could be better predictors of land use changes than phenomena occurring at the farm level.

What this type of study shows is that some parts of the landscape may change greatly within a few years because land is arable and thus may be plowed, farms may be large, and individual farmers close to retirement. On the other hand, if a young farmer takes the farm he may convert low-input grassland with high species diversity (up to 116 species in a 2- to 3-ha meadow — Baudry and

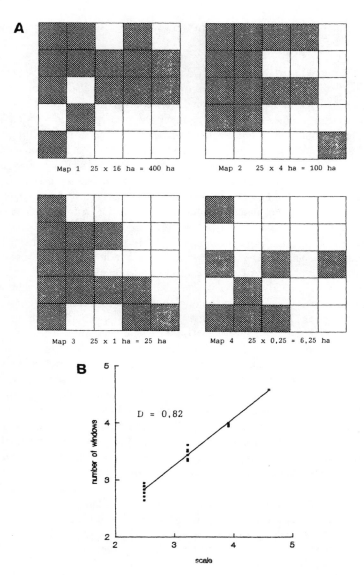

Figure 5. (A) Fractal distribution of ungrazed patches in Sainte-Marguerite-de-Viette. Maps represent the presence of ungrazed patches in the windows. Enlargements are 4 times the upper left corner of the preceding map. Each magnification reveals more details. Map 4 is a 64 times magnification of map 1. (B) The graph demonstrates that the data collected at the 16-ha scale permit an accurate estimate of the number of 0.25-ha cells where ungrazed patches are present. D, the fractal dimension, is the slope of the regression line between the scale and the number of windows with ungrazed patches at each scale (data are log transformed); for coarser scales, several grids were used.

Szujecka, unpublished data) into a corn field within a few weeks. Therefore, he creates a different environment in terms of habitat, water flow, and nutrient cycling. It also appears that nonarable land within crop farms may be used to feed beef cattle, with low stocking rates. Different levels of land use intensity may coexist, not only within the same microregion, but also within the same farm, and landscape patterns are under different levels of control.

These two short studies of land use changes and principal factors of landscape patterns carried out at different scales emphasize the fact that it is difficult to predict changes in the landscape. Predictions of the ecological consequences of those changes are not easy at a landscape and/or regional scale. In the case of a landscape dominated by permanent grassland where meadows are surrounded by hedgerows, two related questions are posed — one as regards farming systems: "Where will they change, under what constraints, and how fast?" — the other as regards ecological systems: "How will changes affect regional species diversity, and what are the rates of change?" More precisely, what are the time and spatial scales at which we are able to make predictions? Do these scales match for ecological and agricultural patterns and processes?

A first methodological problem is the linkage between farming systems and landscape patterns. Because they belong to two non-nested hierarchies, it is impossible to relate a specific farming system to a type of landscape pattern. The method used here, collecting data on the same grid, introduces errors and possibly leads to misleading conclusions. Other methods need to be tested, and other structural variables added. Conway (1988) presents a method for evaluating the importance of structural variables for the determination of farm profit, which could be used to discriminate among the different possible variables. Using more variables would be possible as more farms are studied (if too many variables are used, regarding the number of farms, each farm may appear unique and be an outlier). Variables describing the relationships between farms is essential in a hierarchical approach (Balent, 1989); in a larger sample of farms in Normandy, Laurent (1991) has shown that farm structure is related to extra income and other social variables.

The second problem is that different species, be they plants or animals, behave at different scales in space and time. Fine grain species, e.g., Araneidea (Asselin and Baudry, 1989) see rapid changes among highly unstable habitats, while coarse grain species (e.g., birds of prey) probably see slower, but hardly reversible changes. This interpretation may hold to explain why Middleton and Merriam (1983) found most of the forest species in farm woodlots, as compared to the forest, two centuries after the beginning of agriculture in eastern Ontario, Canada. Only large mammals were absent.

There are also differences between farming systems on the time scales at which they operate. Usually, more intensive systems have rates of behavior faster than low input systems, e.g., in terms of changing crops or cattle breed, the use of fertilizers, and so on.

It is thus necessary to "scale" ecological phenomena under investigation according to the scale of land use changes. It might be that there are so many lags in the response of ecological processes and patterns to changes that it is difficult to relate a structure to an observation. Burel (1992) provides an example of one such lag by showing that carabid distribution may be better explained by landscape structure change over a period of 30 years than by current structure. The ability of a species to move as fast as the changes that occur must be considered when trying to explain its distribution.

Fractal Geometry: A Conceptual Framework to Deal With the Problem of Scale and Prediction

Fractal geometry is a useful tool to measure spatial patterns and to pose hypotheses on their ecological effects as well as their dynamics. This will be seen for ungrazed grassland (land abandonment) and plowing. It may also be used to deal with changes of scale in space and time and, as such, could be used in the future to link different levels of organization and to extrapolate predictions from one scale to another.

Land Abandonment and Plowing

These two types of land cover have a fractal distribution, but with a different D (Table 6). They are therefore fragmented and distributed all over the landscape. The low fractal dimension of ungrazed patches (less than 1, the dimension of a line) reflects the presence of many small patches in meadows. Studies on spiders (Asselin and Baudry, 1989), carabids (Burel, 1991), and passerine birds (Balent and Courtiade, unpublished) show that only the arthropods react to these patches; while they may be too small to affect the distribution of birds that behave at a coarser grain. Morse et al. (1985) have shown the relationships between the fractal dimension of plants and the species diversity of insects found on them. Plowland (D = 1.47) is still fragmented but some large patches exist. D is less than the one for arable land (D = 1.7) because fragmentation of plowed land is due to both fragmentation or arable land and fragmentation of plowing within it.

Sugihara and May (1990) suggest that D may be related to dynamic rates. More fragmented patches would be less stable than compact ones (Sugihara and May computed the fractal dimension of perimeters, their D is higher for fragmented patches; when the computation of D is done on patches, as in this study, D is smaller for fragmented patches). Casual observations suggest that farmers occasionally mow ungrazed patches, which may reconvert quickly (Asselin and Baudry, 1989), whereas plowland is rarely reconverted to grassland. This supports the hypothesis of D as an indicator of stability.

Changing Scale

If the sampled patches are fractal, fine scale information (insect data) can be extrapolated from one scale to another. For example, one can estimate the frequency of a species over a whole landscape by knowing the frequency of patches at a coarse scale. Coarse grain observations give information on fine grain patterns. In this case, patches are assimilated to ecological niches (Kolasa, 1989). Thus, knowing the fractal dimension of ungrazed patches, it is possible to have the number of 0.25-ha cells where they can be found only by counting the number of 16-ha windows where they are present. Of course, this is only a statistical self-similarity and one cannot exactly know where the 0.25-ha cells with ungrazed patches are located, but if the result is general it would render the task of estimating land abandonment easier. Another advantage of fractals is that instead of finding a common scale for all the studies, it may be more efficient to choose, within this range of 0.25/16-ha scales, the scale most convenient for different studies. Then it would be possible to relate the observations made in the different types of patches by using their fractal dimension.

The other point to explore is whether or not rates of changes are fractal. If they are, it will be possible to extrapolate, within the fractal domain, fine scale observations over a short period of time to a longer term and a larger area, taking into account nonlinearities.

CONCLUSION

Changes in landscape, both agricultural and ecological, result from interactions between large-scale constraints (market and species pool) and fine-scale processes (farming and colonization). One of the key problems in estimating the ecological impact of changes in farming systems is extrapolation to regional levels of the fine grain observations usually carried out in ecological and agricultural studies. It is possible that fractal geometry may provide an appropriate procedure, but more investigations are necessary before definite conclusions may be reached. Furthermore, the links between farming activity and landscape patterns have still to be improved.

ACKNOWLEDGMENT

I thank Françoise Burel for comments on earlier versions of the manuscript and Daniel Denis for his help.

REFERENCES

Allen, T. F. H. and Starr, T. B. (1982) *Hierarchy: Perspectives For Ecological Complexity.* University of Chicago Press, IL, 310.
Asselin, A. and Baudry, J. (1989). Les araneides dans un espace agricole en mutation. *Acta Oecolog. Oecol. Appl.* 10:143–156.
Balent, G. (1989). Hierarchical analysis of spatial patterns in pastoral systems. *Etud. Rech. Syst. Agrair. Dev.* 16:187–198.
Baudry, J. (1989). Interactions between agricultural and ecological systems at the landscape level. *Agric. Ecosys. Environ.* 27:119–130.
Baudry, J. (1991). La perception des changements d'utilisation des terres à différentes échelles. *Les Transfers d'Échelle,* C. Mullon, (Ed.), ORSTOM, 425–438.
Baudry, J. (1991). Ecological consequences of land abandonment: role of interactions between environment, society and techniques. Agricultural land abandonment and its role in conservation. INTECOL Semin., Zaaragoza, *Options Mediteranéennes,* A15:13–19.
Bazin, G., Larrere, G. R., De Montard, F. X., Lafarge, M., and Loiseau, P. (1983). Système agraire et pratiques paysannes dans les Monts Dômes. INRA, Paris, 317.
Benzécri, J. P., (Ed.) (1973). *L'Analyse des Données,* Tome 1: la taxinomie. Donod, Paris, 615.
Burel, F. (1992). Effect of landscape structure and dynamics on carabid biodiversity in Brittany France. *Landscape Ecol.*
Burel, F. (1991). Ecological consequences of land abandonment on carabid beetle distribution in two contrasted grassland areas. *Options Méditerranéennes,* A15:111–119.
Burel, F. and Baudry, J. (1990). Structural dynamic of a hedgerow network landscape in Brittany France. *Landscape Ecol.* 4:197–210.
Conway, A. G. (1988). Land in economic models of location — a review. *Agriculture: Socio-Economic Factors in Land Evaluation,* J. M. Boussard (Ed.), Comm. Eur. Commun. Rep. EUR 11269: 67-83.
Denis, D. (1991). Applications de DBase IV pour la construction de bases de données d'utilisation des terres à plusieurs échelles. INRA Working paper (unpublished).
Flaherty, M. and Smit, B. (1982). An assessment of land classification techniques in planning for agricultural land use. *J. Environ. Manage.* 15:323–332.
Golley, F. B. and Ryszkowski, L. (1988). How can agroecology help solve agricultural and environmental problems? *Ecol. Int.,* 16:71-75.
Hubert B. and Girault, N. (Eds.) (1988). De la touffe d'herbe au paysage: troupeaux et territoires, échelles et organizations. INRA Publ., Versailles, 336.
INRA, ENSSAA (Groupe de Recherche) (1977). *Pays, Paysans, Paysages dans les Vosges du Sud,* INRA, Paris, ENSSAA, Dijon, 192.
Iversen, L. R. (1988). Land-use changes in Illinois, USA: The influence of landscape attributes on current and historic land-use. *Landscape Ecol.* 2:45–61.
Kolasa, J. (1989). Ecological systems in hierarchical perspective: breaks in community structure and other consequences. *Ecology,* 70:36–47.
Kullback, S. (1959). *Information Theory and Statistics.* John Wiley & Sons, New York, 215.
Laurent, C. (1991). Place de l'activité agricole dans l'espace, l'emple d'une région agricole de Normandie, le Pays d'Auge. *Econ. Rurale,* 202–203.

Lebeaux, M. O. (1985). ADDAD Association pour le Développement et la Diffusion de l'Analyse des Données Multigraph, 195 pp.
Loehle, C. (1983). The fractal dimension and ecology. *Specul. Sci. Technol.* 6:131–142.
Mandelbrot, B. (1984). *Les objets fractals.* Flammarion, Paris, 203.
Middleton, J. and Merriam, H. G. (1983). Distribution of woodland species in farmland woods. *J. Appl. Ecol.* 20:625–644.
Milne, B. T. (1988). Measuring the fractal geometry of landscapes. *Appl. Math. Comput.* 27:67–76.
Milne, B. T. (1991). Lessons from applying fractal models to landscape patterns. *Quantitative Methods in Landscape Ecology,* M. G. Turner and R. H. Gardiner (Eds.), Springer-Verlag, New York, 197–235.
Morse, D. R., Lawton, J. H., Dodson, M. M. and Williamson, M. H. (1985). Fractal dimension of vegetation and the distribution of arthropod body lengths. *Nature.* 314:731–733.
O'Neill, R. V. (1989). Perspectives in hierarchy and scale. *Perspectives in Ecological Theory.* J. Roughgarden, R. M. May, and S. A. Levin (Eds.), Princeton University Press, Princeton, NJ 140–156.
Phipps, M. (1981). Entropy and community pattern analysis. *J. Theor. Biol.* 93:253–273.
Phipps, M., Baudry, J. and Burel, F. (1986). Dynamique de l'organisation écologique d'un paysage rural: Modalités de la désorganisation dans une zone peri-urbaine. *C.R. Acad. Sci. Paris, T. 303, Sér. III,* 7:263–268.
Sugihara, G. and May, R. M. (1990). Applications of fractal in ecology. *TREE.* 5:79–86.
Turner, M. G. (1990). Spatial and temporal analysis of landscape patterns. *Landscape Ecol.* 4:21–30.
Voss, R. F. (1988). Fractals in nature: from characterization to simulation. *The Science of Fractal Images.* H. Peitgen and D. Saupe (Eds.), Springer-Verlag, New York, 21-70.
Wiens, J. A. (1989). Spatial scaling in ecology. *Functional Ecol.* 3:385–397.
Wilkinson, L. (1988). *SYSTAT: The System for Statistics.* SYGRAPH. SYSTAT Inc., Evanston, IL.
Zonneveld, I. S. and Forman, R. T. T. (Eds.) (1989). *Changing Landscapes: An Ecological Perspective.* Springer-Verlag, New York, 286.

4 Landscape Structure and the Control of Water Runoff

FRANÇOISE BUREL
JACQUES BAUDRY
JEAN-CLAUDE LEFEUVRE

Abstract: We describe how the hedgerow network of Brittany (France) was for the most part designed to control water flow, and identify functional units of sets of hedgerows that channel water and prevent erosion. A case of hedgerow removal leading to erosion is presented.

INTRODUCTION

Rural landscapes in western Europe have changed dramatically over the last few decades (e.g., Agger and Brandt, 1988; Burel and Baudry, 1990). Many uncultivated elements, such as woodlots or hedgerows, have been removed in order to facilitate cultivation. In Brittany (France) there are two main formations of old rocks, i.e., granite and shale, which do not permit penetration of rainfall into the water table. The topography is gently rolling, and the landscape is characterized by a hedgerow network which reached its highest density in the early 1900s (Meyer, 1972). The main ecological function assigned by farmers to this network was to control superficial waterflow (Baudry, 1989). It was not designed nor even managed for other functions, such as the conservation of biodiversity. Changes in the landscape structure and in farming practices have led to disturbance in the control of runoff and, consequently, to an increase in sheet and gully erosion. These phenomena are studied in a hierarchical approach, from the field where physical and agricultural characteristics are described, to the overall landscape whose spatial structure is related to water fluxes.

HEDGEROW NETWORK ORGANIZATION

The control of water flow either for drainage or to prevent flooding has been a constant concern in rural management, especially in regions such as Brittany where climatic and physical conditions generate a significant runoff (Palierne, 1971).

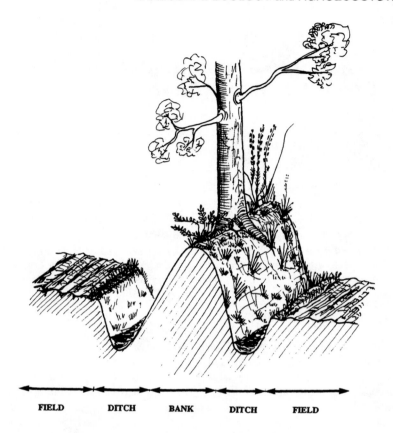

Figure 1. Cross section of a hedgerow, an earthen bank, and a bordering ditch.

Most of the hedgerows have been planted on earthen banks with the earth dug out of one or two parallel ditches (Figure 1). The size and shape of the bands vary geographically, with the mean size being 1.5 m high and 2 m wide (Meynier, 1976). Ditches, which were regularly managed, intercepted the water running down the slopes and directed it toward the main streams.

At the landscape level all the ditches were connected in a functional network, organized according to physical constraints such as topography and soil type. Figure 2 illustrates the distribution of hedgerows and associated ditches along a slope. At the top, the first ditch delimits a plateau separating shallow soils from the deeper ones on the slope. A series of parallel hedgerows oriented perpendicularly are located on the slope. In general, eroded soil particles have accumulated behind these linear elements and created differences in the levels between fields of up to 2 m. In the valleys a ditch and associated hedgerow usually surrounded the flood plain, at the boundary line between mesic and hydromorphic soils. Some trees were planted along the main stream to protect the banks and

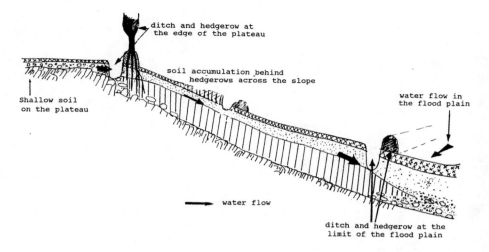

Figure 2. Soil transect along a slope in Brittany and localization of major hedgerows at the edge of the plateau and at the edge of the flood plain.

play a role in the denitrification process. A number of hedgerows parallel to the slope delimit fields and form a closed network.

CONTROL OF WATERFLOW AT THE LANDSCAPE LEVEL

At the field level the ditches control water flow either by stopping water running along the surface or by holding it within the superficial soil layers, or by acting as a channel for water which has been concentrated by the previous elements. Two types of hedgerows may therefore be defined according to their function: those which enhance waterflow, and those which act as barriers.

At the landscape level, hedgerows and ditches are connected in functional units. A functional unit is a set of landscape elements that control an ecological process. Landscape elements are linked together by this process (in the present case — waterflow) which determines landscape organization and function.

In Brittany three types of functional units have been identified (Figure 3).

The first, Unit A, controls the water flow on the slopes, where a ditch across the slope stops water which accumulates within the superficial soil layers in a small depression, whereas another channels the water down the slope.

The second, Unit B, controls the water flow from the plateau, because a ditch at the plateau edge stops all the water running along the plateau. The quantity of water may be very important even if the slopes are shallow because the plateaus are long and the water remains near the surface on shallow soils. Another ditch, as in (A), channels the water down the slope.

The third, Unit C, diminishes the flooding intensity in stream corridors. The ditches which surround the flood plain prevent it from receiving all the water coming from the upper fields, thus limiting flooding.

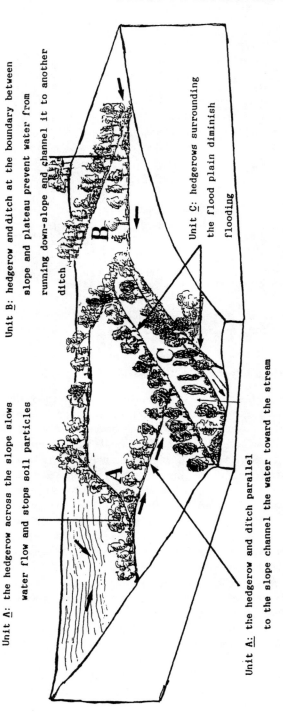

Figure 3. Diagrammatic representation of functional units controlling water fluxes in a hedgerow network landscape. Each functional unit is composed of a set of hedgerows and their associated ditches, some stop water, others channel it down slope. (Adapted from Burel, F. and Baudry, J. [1990]. *Landscape Ecol.*, 4:197–210. With permission.)

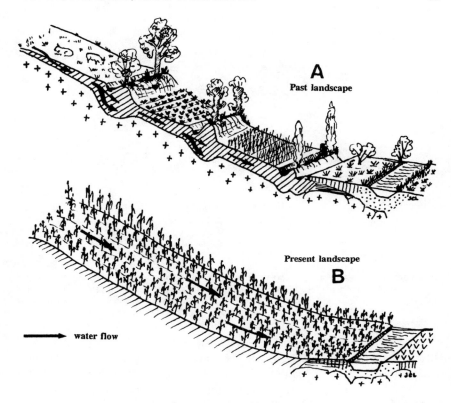

Figure 4. Changes in agricultural landscapes during the last few decades in Brittany. (A) Past landscape; (B) present landscape. Hedgerows perpendicular to the slope have been removed in order to enlarge fields, meadows have been converted into plough fields, corn is cultivated on huge areas whatever the quality of the soil. In the valley, wetlands have been drained.

Water control is dependent upon this integrated set of hedgerows, but when the network is broken up, their efficiency in water flow control is low and may even enhance gully erosion because water is concentrated and not channeled into a connected ditch.

EFFECTS OF LANDSCAPE CHANGES ON WATER FLOW

Changes in agricultural practices, such as the plowing of permanent meadows, the use of new machinery, or the introduction of new crops such as corn for silage, lead to fields being enlarged and to becoming geometric in shape. The new structure of the landscape, created either by individual farmers or by a municipality within a reallotment program, rarely takes into account any ecological processes (Lefeuvre, 1979). Many hedgerows have thus been removed with no consideration for their contribution to the functional units (Figure 4).

46 LANDSCAPE ECOLOGY and AGROECOSYSTEMS

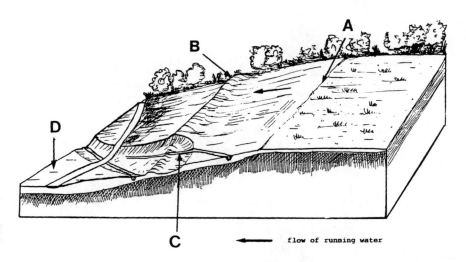

Figure 5. Example of an erosion phenomenon at the landscape level. After hedgerow removal, water runs freely in fields. (A) Ancient ditch and hedgerow at the plateau; (B) topographic coincidence at the location of a previous hedgerow speeds up water flow; (C) gully erosion; and (D) accumulation of soil particles.

In many cases, erosion has occurred where the functional units have been altered (Baudry et al., 1987; Burel and Baudry, 1989). At the field level the risk of erosion depends upon the slope angle, slope length texture of the superficial soil layer, the quantity of water coming into the field either by rainfall or by runoff from upper fields, the nature of the crop, and the farming practices. At the landscape level all the structures which control water flow will also diminish sheet erosion.

The change in landscape structure which enhances erosion risk is enlargement of field size, leading to increased slope length and the breakdown of functional units. An example of such a process is shown in Figure 5 where all the ditches and hedgerows on a slope have been removed. The quantity of water running over the soil is sufficient to erode soil particles and create gullies. Where there were previously hedgerows, there was a topographic coincidence which accelerated the speed of water flow. There was a small track that was previously bordered by two parallel hedgerows which stopped water coming from the slope by acting as a temporary stream. Now, with no hedgerows nor ditches, the fluxes of water and soil particles run directly across the road and accumulate in the lower fields.

If erosion is to be prevented when the fields may be enlarged then the functional units identified in the landscape must be kept in place. In the example above, a hedgerow and ditch at the edge of the plateau and another bordering the road and connected to the existing ones, would be needed to protect the soils from erosion.

CONCLUSION

In Brittany there is a hedgerow network landscape in which many hedgerows and associated ditches play a major role in the control of water flow. At the field level they stop or control fluxes, whereas at the landscape level the sets of connected elements control the flow. Landscape design is a tool to prevent sheet erosion and it will be progressively taken into account since agricultural practices alone are not sufficient to ensure soil protection (Boiffin et al., 1986).

ACKNOWLEDGMENT

We thank Maurizio Paoletti for inviting us to present our work at the INTECOL Congress, and the Ministries of Environment and Agriculture for financial support.

REFERENCES

Agger, P. and Brandt, J. (1988). Dynamics of small biotopes in Danish agricultural landscapes. *Landscape Ecol.*, 1:227–240.

Baudry, J., Trotel, M. C., Burel, F. and Asselin, A. (1988). L'érosion des terres agricoles dans le Massif Armoricain Ministères de l'Environment et de l'Agriculture, CERESA Imprimerie Nationale, 33.

Baudry, J. (1989). Interactions between agricultural and ecological systems at the landscape level. *Agric. Ecosyst. Environ.*, 27:119–130.

Boiffin, J., Papy, F. and Eimberck, M. (1986). Influence des systèmes de culture sur les risques d'érosion par ruisselement concentré. I: Analyse des conditions de déclenchement de l'érosion. *Agronomie*, 8:663–673.

Burel, F. and Baudry, J. (1989). Hedgerow network patterns and processes in France. *Changing Landscapes: An Ecological Perspective*, I. S. Zonneveld and R. T. T. Forman (Eds.), Springer-Verlag, New York, 99–120.

Burel, F. and Baudry, J. (1990). Structural dynamics of a hedgerow network landscape in Brittany France. *Landscape Ecol.*, 4:197–210.

Lefeuvre, J. C. (1979). Les études scientifiques, un préalable indispensable à la restructuration foncière et à l'aménagement des zones bocagères. In Lefeuvre, J. C., *Les Connaissances Scientifiques Écologiques, le Développement*. J. C. Lefeuvre, G. Long and G. Ricou (Eds). Ecologie et Développpment. CNRS, 169–192.

Meyer, J. (1972). L'évolution des idées sur le bocage en Bretagne. *La Pensée Géographique Française Contemporaine*. Presses Universitaires de Bretagne, 453–467.

Meynier, A. (1976). Typologie et chronologie du bocage. *Les Bocages: Histoire Écologie Économie*. CNRS, ENSA, Université de Rennes, 65–68.

Palierne, J. M. (1971). Milieu naturel et paysages agraires. Cahiers Nantais, *Inst. Géogr. d'Aménage. Rég. Nantes*, 3:61–143.

5 Agricultural Transport and Landscape Ecological Patterns

H. GULINCK
F. PAUWELS

Abstract: *Agricultural transport patterns may change dramatically in relation to farming activities, mechanization, field pattern, road structure and quality, road management, and other factors. These patterns and the associated functional processes are the structural basis for a number of landscape-ecological patterns such as material distribution, linear habitats, and runoff. The stability analysis of these patterns can be related to the modeling or changeable farming activities in the landscape. A case study for a site in central Belgium is given.*

INTRODUCTION

In current agroecosystem research, arguments of stability, sustainability, and other goals are increasingly included together with productivity (Conway, 1987; Marten, 1988), which makes an extended approach of agriculture necessary in relation to the environment. There is no doubt that "landscape" is a valuable entry, not only as a label to a spatial-hierarchical level in agroecosystem research, but also to the planning and management of agriculture in its relation to the overall environment. In this respect, agricultural and rural roads may be considered as a significant category of landscape structure, which not only spatially, but also conceptually and functionally, provides a link between agricultural systems, agroecosystems, and the overall landscape. As such, they are physical frameworks which function as a matrix. An understanding of the impacts of agricultural activity in the study of landscape dynamics is crucial (Baudry, 1989), and agricultural transport is a key factor in this relationship. However, little attention is given in agroecology to transport and roads, since they are not a part of the ecological factors that contribute directly to biological production.

Given a distribution in time and place of land use, and a farming management system with a certain use of the available transport facilities, it is the distribution and the type of the roads that largely determines the patterns, intensity, and character of agricultural transport. On the other hand, changing needs for agricultural transport induce alterations in circulation and road patterns. Road patterns and transport are both closely associated with, and codeterminants of, landscape-ecological patterns and processes.

REVIEW OF THE ROLES OF RURAL ROADS

The role of rural roads has been intensively studied in different contexts. Since Von Thünen, geographers have been studying the role of distance in causal relation to various aspects of farming systems. De Lisle (1978) provides an interesting example, taking into consideration details of farm layout, rather than aggregated bookkeeping accounts. Transport is one of the major factors that link urban and rural systems.

A large body of literature exists on the role of roads in developing countries, such as their impacts on poverty alleviation, village development, mobility, farming systems, migration, agroindustries, etc. (Howe and Richards, 1984; Amadi, 1988). Models have been developed for predicting the impacts of road development on agricultural production (Beenhakker and Chammari, 1979).

Roads are an important aspect in watershed management (Gil, 1985), land consolidation, and land reclamation programs (Steiner and Van Lier, 1984) and are intimately linked with agricultural mechanization and the economy of scale. Especially in The Netherlands, considerable efforts have been devoted to the development of operational design and impact models for rural and farming road networks, linkages, and circulation (Jaarsma, 1988).

In agricultural systems and agroecosystem research, the functional roles of roads are subjugated in comparison with the importance given to the central operational and ecological units such as crops, soil and water components, fertilizers, capital, and labor. In mechanized, rational farming systems, farming traffic and transport represent only a small fraction of energy input, in comparison with fertilizers and field cropping practices (Pimentel et al., 1975). On scales above the farming system or landscape level, the transport of energy and of agricultural outputs between regions, nations, and continents is very important, not only in economic but also in environmental terms. The impacts of these processes at a global, continental, and regional scale are also manifest at local or landscape scale, and here the concept of landscape ecology can play an important role in the study of agroecosystems.

Rural roads can be subdivided into hierarchical levels, characterized not only by dimensions such as length, width, and construction, but, more importantly, by the rank of geographical elements they link, or by the amount of traffic and transport they carry. A general hierarchy can already be traced in the distinction, blurred as it is, between rural, agricultural, and farm roads. The latter, opening up fields or at most linking farmsteads and small villages, and often unpaved, are called "minor rural roads" (Michels and Jaarsma, 1988). It is mainly towards this category that the attention of this paper is directed.

In many cases, farming roads and tracks exclude nonagricultural traffic and are to be considered as a lower hierarchical category, but are important because they are the "capillaries" which open up large tracts of agricultural land that still covers the most of the anthropomorphic landscapes in the world.

THE LANDSCAPE-ECOLOGICAL ROLE OF AGRICULTURAL ROADS AND TRANSPORT

Processes in the landscape are influenced to a certain degree by tangible landscape structures and components, such as topography, patch size, and vegetation. Some farming activities are to be considered partly as intangible or unpredictable (e.g., farming perception, market conditions). Nevertheless, they generate and influence important landscape patterns and processes such as crop and cattle distribution, field consolidation, land abandonment, erosion patterns, and nutrient runoff. Agricultural transport, the combined result of these farming activities and of tangible landscape structures, in turn, influences the rural landscape.

The landscape-ecological role of agricultural roads and traffic cannot be neglected. Road verges are ecological features that have received increasing attention as corridors and as reservoirs or refuges for organisms. The rural road pattern together with the field pattern, determines to a great extent the mesh size of the anthropomorphic, agricultural landscape. In general terms, these roads are important elements and patterns in the chorological dimension of agroecosystems (Gulinck, 1986), intimately linked to field patterns and to various landscape-ecological processes.

At least four principles of interaction between agricultural roads and transport and landscape-ecological processes and patterns can be recognized:

Association — There is a congruency, a colinearity of farming transport patterns with, for example, linear biotopes or with lines of concentrated runoff.

Distribution — Organisms and material such as fertilizers, soil matter, and loads of transport are distributed over the landscape, not only by man, but also by all kinds of organisms that use the roads and their verges as corridors.

Separation — Roads and transport can act as barriers against flows of organisms and material. A road can be simultaneously a separator and a distributor.

Disturbance — Not only by the traffic itself and the noise it produces, but also, for instance, by the dissemination of disturbing objects and physical pressure can agricultural transport act as a disturber.

In this paper it is primarily the first category, association, that is further developed as a case study.

FACTORS THAT DETERMINE AGRICULTURAL TRANSPORT

Because agricultural transport is indisputably an exponent of the dynamics of the rural landscape, it is important to recognize the factors by which it is determined. The following groups of factors can be distinguished:

1. Land use
   A. Location and distribution of farms and fields
   B. Size of fields
   C. Farming type, the crops grown, and/or the animals bred

2. Transport facilities
   A. Mechanization on level, transport equipment
   B. Quality of roads
   C. Redundancy of roads, distribution, location and density
3. Soil and crop management
   A. Conventional vs. no tillage
   B. Need for fertilization, pesticide application, etc.
4. Factors external to the farm

A CASE STUDY IN CENTRAL BELGIUM

To illustrate the importance of agricultural transport in the landscape and to show how it can be altered to a very large degree by changing farming activities, road patterns, and quality, a simple transport model is applied in three scenarios on a rural area in central Belgium.

In a preliminary study (Van Meldert et al., 1991) a random nested sampling selection of 17 quadrants of 6.6 km^2 each, was made within a 1270-km^2 rectangular section of the central Belgian loess area. Within this area, an estimation of approximately 4 km/km^2 of minor rural roads was made. About 20% of these roads are sunken, with a depth varying between 0.5 and 15 m below the level of adjacent fields. From this survey, a first picture of "association" was derived: over 60% of the verges of these sunken roads carry woody vegetation, and many of the verges are to be considered as valuable natural biotopes.

The functional relationship of the traditional minor road pattern in relation to agriculture is changing rapidly. Over 15% of the sunken roads also are used for nonagricultural traffic, whereas about 20% seem to have lost all motorized traffic during recent years. Other roads have an increased transport intensity and frequency, and in recent years they have been surfaced and widened for these reasons, with loss of associated vegetation.

A more elaborate analysis of association, distribution, separation, and disturbance, as well as the analysis of the factors of agricultural dynamics, has been started for a small sample area of about 1000 ha near the village Hoegaarden (Figure 1). It forms a landscape in the loessbelt of Belgium characterized by fragmented farmland, a dense network of unsurfaced sunken farmroads, and a high density of linear vegetation elements which are closely linked with the road pattern (Figure 2). Using a digital relief model and a catchment and surface runoff model, potential lines of concentrated flow were derived and matched with the pattern of farming roads (Figure 3). Many sunken roads coincide with such lines.

Rapid changes take place in the agricultural landscape. An aging farming population is leading to the formation of fewer large farming units and a thorough restructuring of activity patterns. This area will be subject to a land consolidation program in the coming years. A careful assessment of the potential impacts of the scenarios of evolution and consolidation on the landscape-ecological patterns and processes is necessary.

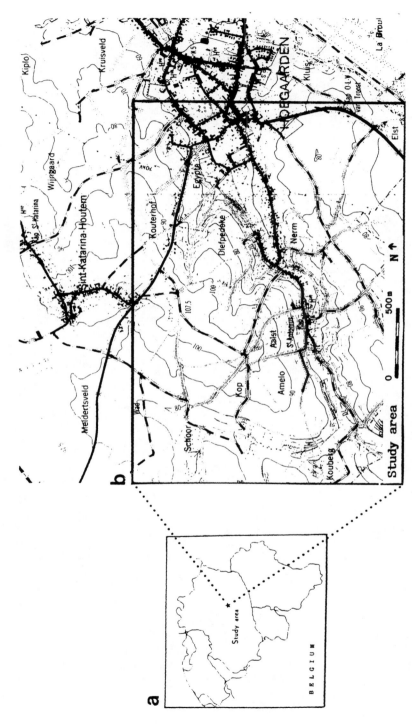

Figure 1. (a) Localization of Hoegaarden in Belgium. (b) The study area near the village of Hoegaarden; note the high density of sunken roads (▨) (N.G.I. topographical map).

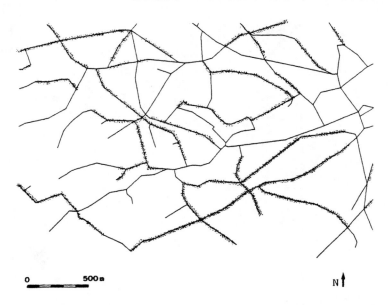

Figure 2. Linear vegetation elements of high local and regional value (fauna and flora) in direct association with roads. (~~~~) Road associated with biotopes of high value.

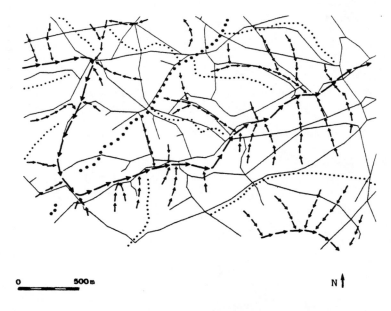

Figure 3. Hydrology, generated by a digital relief model. (→→) runoff lines; catchment area >20 ha; (····) 1st and (· ·) 2nd order catchment area boundaries.

Three scenarios were developed in order to generate a simulation of the internal agricultural transport pattern and intensity in the Hoegaarden landscape.

Current Situation

To generate this pattern (Figure 4a) the following data were used:

- Present location and area of fields used by three farms which represent about 20% of the total field area. To obtain this, combined cadastral data and data from a direct inquiry were used.
- Actual location of farmsteads — farmstead number 1 lies just outside the area shown in Figure 4a.
- Mean annual frequencies per hectare of farmstead-field transport for different crops (Figure 5), obtained by inquiries and observations.
- Actual road pattern and quality, determined through visual judgment.

For road quality, the practicability for agricultural transport of each road segment is represented by its impedance, given in an ordinal scale; attributes of this quality are the character of the surfacing, road width, slope, and curvature. The impedance of a road can be considered as the "difficulty or the relative unwillingness" of using it (Blunden and Black, 1984). Roads not practicable for normal agricultural transport received an impedance of ∞; roads with a hard surface and a carriageway of more than 3 m received an impedance of 1; the remaining, difficultly practicable roads received an impedance of 2. We can assume that these impedance levels correspond with factors of reduction in speed, increase in transport costs, or of apparent increase in road length.

To determine the field-farmstead path, the shortest route is taken, using a weighting factor which lengthens a road with a factor that equals its impedance level. The intensity is determined by the area of the target field. To simplify the calculations, an annual transport journey frequency of 43 per ha is incorporated — a number taken as the mean for the two almost exclusively grown crops of the region, wheat and sugar beets.

The scenario obtained in this simple way was found to match the reality closely as far as the relative use of each link of the road network is concerned.

Road Improvement

The second scenario (shown in Figure 4b) assumed an overall road improvement so that all existing roads received impedance 1. The farmers use the shortest paths from and to their fields. All the other data of the previous scenario remain unchanged.

Field Consolidation and Extension

The transport pattern in Figure 4c is the result of a realistic reallotment and consolidation scheme, which centralizes the fields closer to the farmsteads.

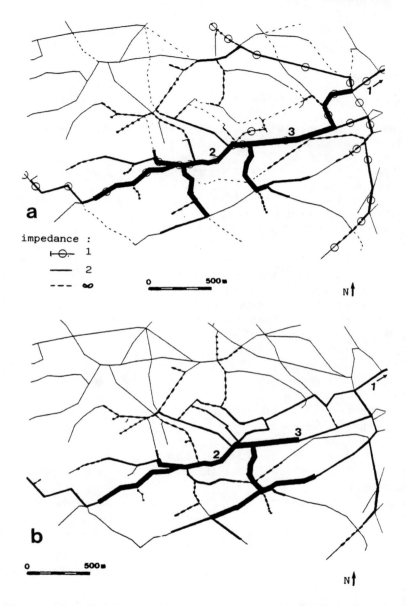

Figure 4. Annual journal frequency: (━━) >365; (───) >52 and <365; (─ ─) >12 and <52; and (┈┈) >0 and <12. 1, 2, and 3 are farmsteads. (a) Scenario I: actual situation. (b) Scenario II: road improvement. (c) Scenario III: field consolidation.

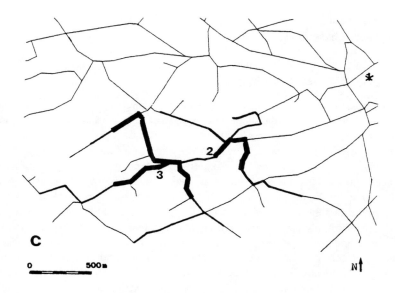

Figure 4c.

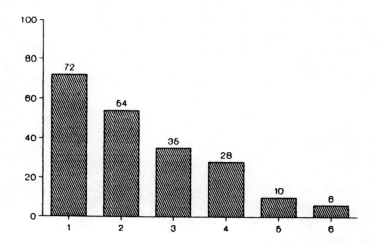

Figure 5. Mean annual journey frequencies per hectare of farmstead — field transport for different crops: (1) potatoes; (2) sugar beets; (3) wheat and barley; (4) maize; (5) pasture; and (6) aftercrop.

Furthermore, farm number 1 is supposed to have stopped all activities and the biggest farm, number 2, increased its total field area by taking over the fields of farm 1. Farm 2 is also relocated outside the village center to decrease the intensity of heavy agricultural transport in front of people's doorsteps. Some roads are abandoned, a new road is created, and road quality is improved so that all roads have impedance 1.

DISCUSSION AND CONCLUSIONS

A comparison of the scenarios depicted in Figure 4 gives an impression of the kind of changes that may take place in transport pattern and intensity caused by respective road improvement, field consolidation, and farm relocation. A comparison of the generated patterns with important landscape-ecological patterns such as shown in Figures 2 and 3 helps to understand possible environmental impacts. An increase in transport on a road with an important runoff function may lead to problems downstream if, as a consequence of increased traffic, more erosion of the road surface is induced, or if hard surfacing of the road is applied without adequate control of runoff. An increase in transport in a road associated with an important vegetation corridor may lead to the degradation or removal of this corridor. The other extreme, the abandonment of a road by farming traffic, may lead to its entire removal within a short period.

The spatial and transport models applied here are on a low level of sophistication and should be expanded through the sampling of more farms, the quantification of more transport factors, and the validation through monitoring of actual transport and the development of submodels. Some of these submodels should be statistical rather than deterministic, such as the relation of low-grade roads, or the decisions on road surfacing or verge vegetation cutting in relation to transport intensity, etc. The application of network techniques in a GIS environment should help the development of circulation activity scenarios considerably.

ACKNOWLEDGMENT

The authors would like to thank the referee for his constructive remarks.

REFERENCES

Amadi, B. C. (1988). The impact of rural road construction on agricultural development: an empirical study of Anambra state in Nigeria. *Agric. Syst.*, 27:1–9.

Baudry, J. (1989). Interactions between agricultural and ecological systems at the landscape level. *Agric. Ecosyst. Environ.*, 27:119–130.

Beenhakker, H. L. and Chammari, A. (1979). Identification and appraisal of rural road projects. World Bank Staff Working Paper No. 362, 74.

Blunden, W. R. and Black, J. A. (1984). *The Land-use/Transport System*. Pergamon Press, Elmsford, NY, 250.

Conway, G. R. (1987). The properties of agroecosystems. *Agric. Syst.*, 24:95–117.

De Lisle, D. de G. (1978). Effects of distance internal to the farm: a challenging subject for North American Geographers. *Prof. Geogr.*, 30(3):278–288.

Gil, N. (1985). Watershed Development With Special Reference to Soil and Water Conservation. FAO Bulletin 44, Rome, 257.

Gulinck, H. (1986). Landscape ecological aspects of agroecosystems. *Agric. Ecosyst. Environ.*, 16:79–86.

Howe, J. and Richards, P. (Eds.) (1984). *Rural Roads and Poverty Alleviation*. Intermediate Technology Publications, London, 191.

Jaarsma, C. F. (1988). Het verkeersmodel als hulpmiddel voor de verkeersplanning in een landinrichtingsproject (in Dutch). *Cultuurtech. Tijdsch.*, 27(5):319–329.

Marten, G. G. (1988). Productivity, stability, sustainability, equitability and autonomy as properties for agroecosystems assessment. *Agric. Syst.*, 26:291–316.

Michels, Th. and Jaarsma, C. F. (Eds.) (1989). Minor Rural Roads. Planning, Design and Evaluation. Proc. Eur. Workshop, Wageningen, October, 1987, 198.

Pimentel, D., Dritschilo, W., Erummel, J. and Eutzman, J. (1975). Energy and land constraints in food protein production. *Science*, 190:754–761.

Steiner, F. and Van Lier, H. (Eds.) (1984). Land conservation and development. *Developments in Landscape Management and Urban Planning*, 6B, Elsevier, New York, 481.

Van Meldert, M., Gulinck, H. and Pauwels, F. (1991). Holle wegen in de leemstreek: verspreiding en kenmerken (in Dutch). *Groenkontakt*, 17(1):47–52.

6 Effect of Habitat Barriers on Animal Populations and Communities in Heterogeneous Landscapes

K. DOBROWOLSKI
A. BANACH
A. KOZAKIEWICZ
M. KOZAKIEWICZ

Abstract: *The problem of spatial differentiation of a landscape is analyzed in terms of environmental barriers within it. Based on many examples, the effect of environmental barriers on the structure of bird and rodent populations and communities is discussed. Environmental barriers that act at the level of individual animals are easy to cross and do not fragment the population or community. They can, however, limit the frequency of movements of individuals. They also modify the way of space utilization by influencing the shape and size of the home ranges or territories of individuals. Environmental barriers within a metapopulation disrupt the continuity of the population. Such barriers strongly limit the movements of individuals or species that can cross them. They can therefore influence the ecological structure and dynamics of individual subpopulations or communities. Both types of environmental barriers can affect interspecific interactions. At the ecotones, the presence of one species can act as an additional barrier, limiting movements of individuals of another species.*

INTRODUCTION

It is generally postulated that the degree of spatial heterogeneity of a landscape is one of the factors determining the structure and functioning of animal populations and communities. A homogenous landscape creates identical living conditions for individuals throughout its area. So, the probability of occurrence of different species should be constant at any point in such a landscape, and individual populations should show identical characteristics, that is, a constant density and structure, throughout the area (Figure 1A).

Landscape homogeneity can be disrupted by barriers. Such a system can be considered as homogenous, but divided into parts (Figure 1B). A heterogenous landscape consists of patches characterized by different quality, size, and shape. It may be expected that the various patches of a heterogenous landscape will be

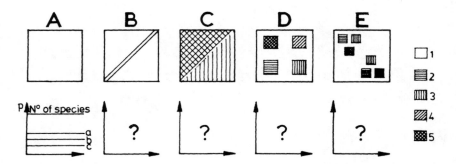

Figure 1. Types of landscape heterogeneity vs. the distribution of animal populations and communities. p: Probability of occurrence of animal populations and communities with distinct characteristics; a–c: population characteristics; 1–5: patches of different habitats; (A) homogenous landscape; (B) homogenous landscape crossed by barrier; and (C–E) heterogenous landscapes.

inhabited by different animal populations and communities, or their parts. This differentiation can be an effect of distinct biotic and abiotic factors in different habitat patches. Boundaries between individual patches, and occasionally between the whole system of patches, can function as natural environmental barriers to animals (Figures 1C, D, and E).

The degree of isolation of population or community parts in a landscape fragmented by barriers depends on the types of the barriers and their effectiveness. For example, population parts separated by a narrow road, or a thin strip of a meadow, crop field, or by small stream will be less isolated than when separated by a highway, broad belt of crop fields, or a large river (Oxley et al., 1974; Mader, 1984; Merriam et al., 1989; and others). Thus, it seems that some of the basic factors determining the degree of spatial heterogeneity of the landscape are the number, type, distribution, and effectiveness of environmental barriers.

The effect of barriers on the functioning of animal populations and communities also depends on the biology of individual species. This point can be illustrated by the effect of the same barrier, which in this case is a road crossing a forest, on two rodent species — the bank vole, *Clethrionomys glareolus* and the yellow-necked mouse, *Apodemus flavicollis*. This road limited the movements of voles, but it did not affect the movements of mice (Figure 2) (Bakowski and Kozakiewicz, 1988). Thus, the degree of landscape heterogeneity should always be defined in relation to individual species or communities (groups of species).

With respect to the scale of their effects, environmental barriers can be classified into: (1) the barriers affecting individuals but not discontinuing the population, (2) the barriers within the boundaries of a metapopulation (*sensu* Levins, 1970), separating subpopulations but not precluding movements of animals, and (3) the barriers totally isolating metapopulations (Figure 3). Only the first two types of barriers are analyzed in this chapter.

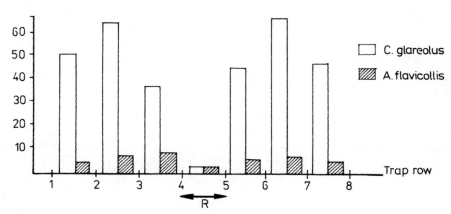

Figure 2. Number of movements of bank voles *(Clethrionomys glareolus)* and yellow-necked mice *(Apodemus flavicollis)* between neighboring trap rows; R: road. (After Bakowski, C. and Kozakiewicz, M. [1988], *Acta Theriol.*, 33:345–353. With permission.)

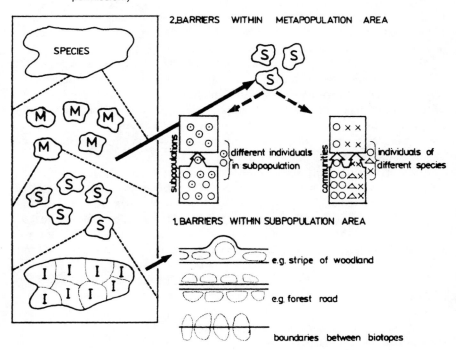

Figure 3. Types of environmental barriers and their effects on animal populations and communities at different spatial scales. Species — area occupied by species; M — area occupied by a metapopulation; S — area occupied by a subpopulation; and I — area occupied by an individual. (After Merriam, G. [1988], *Landscape Ecol.*, 2:227–235.)

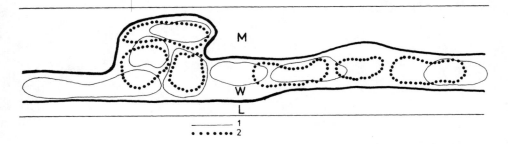

Figure 4. Distribution of breeding territories of the thrush nightingale *(Luscinia luscinia)* (1), and the chaffinch *(Fringilla coelebs)* (2), in a narrow belt of wood at Lake Sniardwy. M — meadow. W — wood belt, and L — lake. (After Rybacki, L. [1980], M.Sc. thesis, Warsaw University. With permission.)

BARRIERS AFFECTING INDIVIDUALS BUT NOT FRAGMENTING THE POPULATION

These barriers are relatively easy to cross, thus they do not disrupt the continuity of the population. They act at the level of individual animals and modify the way space is utilized by individuals by influencing, for example, the size and shape of individual home ranges and territories (Figure 3).

An example of the effect of barriers on the shape of breeding territories is provided by some birds inhabiting narrow belts of woodland and scrub by a lake (Rybacki, 1980). Territories of the thrush nightingale *(Luscinia luscinia)* and the chaffinch *(Fringilla coelebs)* were elliptical in narrower sections of woodlots and almost circular in wider sections (Figure 4). This variation was caused by environmental barriers limiting the occurrence of the breeding population to a small habitat patch.

A linear barrier, e.g., a road, ditch, or side of a forest can limit contacts between individuals living on both sides of it, without precluding contacts between the separated parts of the population. This can be illustrated by a population of bank vole *(C. glareolus)* inhabiting a pine forest crossed by a 5-m-wide road. The two parts of the population were in touch due to individuals passing across the road. However, the number of individuals crossing the road was significantly lower than the number of individuals covering the same distance in both parts of the forest not divided by any barrier (Figure 2) (Bakowski and Kozakiewicz, 1988).

Similar results were obtained for a population of the white-footed mouse *(Peromyscus leucopus)* in heterogenous forests bisected by roads. The movements of animals in the study plots were analyzed and it was found that the road did not isolate the two parts of the population but limited the number of contacts between them (Merriam et al., 1989). It may be suggested that in both these cases (homogenous and heterogenous habitats) the reduction of contacts among individuals inhabiting the two parts of the forest resulted from the separation of

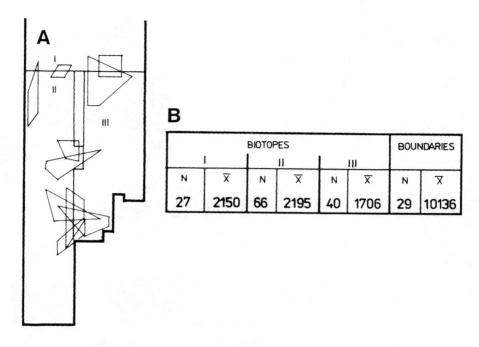

Figure 5. Examples of home ranges of bank voles *(Clethrionomys glareolus)* living at boundaries between habitats (A) and mean sizes of home ranges (in square meters) in different habitats and at their boundaries (B): I — willow thicket, II — alderwood, and III — pine wood. (After Banach, A., [1988], *Acta Theriol.*, 33:87–102. With permission.)

individual home ranges. Presumably, the plot shape and the degree of overlapping along the road differed from those deep in the forest. The road-forest boundary determined the boundaries of individual home ranges, without limiting the area occupied by the population.

Environmental barriers modifying individual home ranges (territories) can be formed by the boundaries between habitat patches of different qualities, but occupied by a population in a continuous way (Figure 3). In the case of the bank vole population, it has been found that home ranges of individuals occupying boundary areas were larger than those of individuals living within habitat patches (Figure 5) (Banach, 1988).

BARRIERS WITHIN METAPOPULATIONS SEPARATING SUBPOPULATIONS

Barriers of this type fragment the total population, limiting its occurrence only to suitable habitat patches. Movement of individuals among patches integrates the system of such subpopulations into a metapopulation *sensu* Levins (1970). The isolation of individual subpopulations increases with the growing barrier

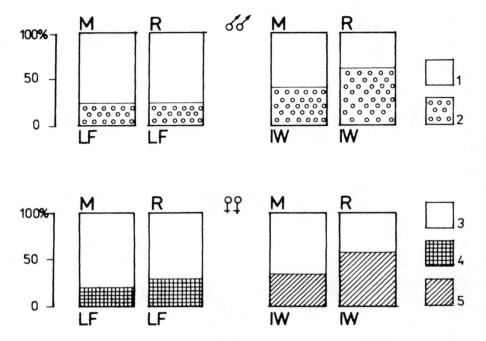

Figure 6. Structure of sexual activity in bank voles *(Clethrionomys glareolus)* recolonizing a large forest (LF) and a small woodlot isolated by meadow (IW): M — transient individuals (newly marked); R — residents; 1 — males sexually inactive; 2 — males sexually active; 3 — females sexually inactive; 4 — reproducing females (pregnant and lactating); and 5 — females sexually active but not reproducing. (Modified from Kozakiewicz, M. and Jurasińska, E. [1989], *Holarctic Ecol.,* 12:106–111. With permission.)

effect. Their ecological distinctiveness also increases (Kozakiewicz and Szacki, 1987).

It seems that one of the factors accounting for a distinct ecological character of individual subpopulations is a "filtration" of some categories of individuals through barriers (Figure 3). An example is provided by population studies of the bank vole. An ecological gap was created by removing all the rodents from a small woodlot surrounded by a meadow and from a plot located within a large forest. The recolonization of these two areas was then observed (Kozakiewicz and Jurasińska, 1989). Bank voles crossing the meadow and recolonizing the isolated woodlot differed from those recolonizing the plot in the forest which had no barrier to cross. The former comprised a higher proportion of sexually active individuals, but reproducing females were absent (Figure 6). Their mean body weight was also lower than that of the bank voles recolonizing the nonisolated forest plot. Thus, some categories of individuals did not cross the environmental barrier.

"Filtration" of some categories of individuals through ecological barriers was also found by Middleton and Merriam (1981) for a whitefooted mouse *(P. leu-*

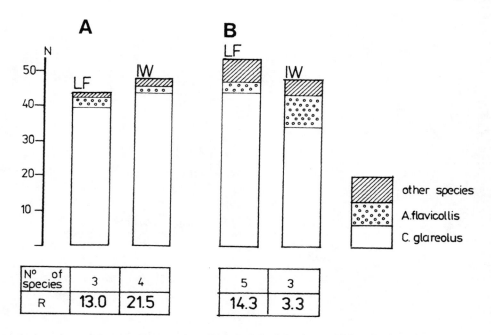

Figure 7. Species composition and numbers of individuals in a small mammal community occupying a large forest (LF) and a small woodlot isolated by a meadow (IW) prior to a removal experiment (A), and in the fourth week of recolonization (B): R — ratio of the number of bank voles *(Clethrionomys glareolus)* to the number of yellow-necked mice *(Apodemus flavicollis)*. (Modified from Kozakiewicz, M. and Jurasińska, E. [1989], *Holarctic Ecol.*, 12:106–111. With permission.)

copus) population and by Hansson (1987) for a bank vole *(C. glareolus)* population.

ENVIRONMENTAL BARRIERS IN RELATION TO ANIMAL COMMUNITIES

As with a single population, individual species can also undergo selective effects of environmental barriers (Figure 3) and this may account, in part, for a distinct character of animal communities in various patches of the same habitat type. This may be illustrated by the removal experiment quoted above and the subsequent recolonization by rodents of the isolated woodlot surrounded by a meadow barrier and nonisolated forest plot. This meadow created a barrier to bank voles *(C. glareolus)* but it did not limit the movements of other rodents as much, e.g., yellow-necked mice *(A. flavicollis)*. As a result, the proportion of individual species in the community after recolonization in the isolated woodlot differed from that in the nonisolated forest plot (Figure 7, Kozakiewicz and Jurasińska, 1989).

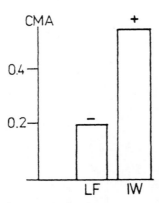

Figure 8. Values of Cole's measure of association (CMA) for a pair of species: bank vole *(Clethrionomys glareolus)* and yellow-necked mouse *(Apodemus flavicollis)* inhabiting a large forest (LF) and a small woodlot isolated by a meadow (IW): (+) — statistically significant tendency to spatial co-occurrence of the two species, (−) — no tendency to co-occurrence. (Modified from Kozakiewicz, M., Kozakiewicz, A. and Banach, A. [1987], *Acad. Pol. Sci. Cl. II*, 35:181–188. With permission.)

Environmental barriers can also influence the spatial distribution of species. For example, bank voles *C. glareolus* and striped field mice *A. agrarius* showed a random distribution with respect to each other at trapping stations in a large forest without barriers, whereas they tended to co-occur in space in the same habitat type but separated by a meadow (Figure 8, Kozakiewicz et al., 1987).

When analyzing the effects of interspecific interactions, it is worth noting that in some cases they may account for additional barriers in the habitat. This can be illustrated by the situation in a small rodent community inhabiting the lake shore zone, which is a diverse habitat with a zonal distribution of different plant communities. Two rodent species were distributed in two different parts of this habitat: bank voles *(C. glareolus)* occupied dry areas, whereas striped field mice *(A. agrarius)* were found in patches close to the edge of the lake. It was shown that the space allocation between these two species was not due only to the effect of differences in habitat preferences. When bank voles *(C. glareolus)* were removed from the dry area, it was very quickly colonized by striped field mice *(A. agrarius)* (Kozakiewicz, 1987). A similar effect was observed at the forest meadow interface. A narrow belt covered with shrubs was occupied only by root vole *(Microtus oeconomus)*. After the removal of this species, bank voles *(C. glareolus)* moved into this area from the adjacent forest (Banach, unpublished).

Another example is provided by two congeneric birds, the great reed warbler *(Acrocephalus arundinaceus)* and the reed warbler *(A. scirpaceus)*. Both species are territorial and typically build their nests in reedbeds, using three or four stems as nest supports. Reed warblers *(A. scirpaceus)* can also nest in shrubs growing near reedbeds. There is therefore likely to be an effect of competition with the great reed warbler *(A. arundinaceus)*. The latter species arrives a little earlier from its winter quarters, and establishes breeding territories in the optimum habitat for both the species, that is, in reedbeds. Reed warblers *(A. scirpaceus)* arrive later and nest in reedbeds only in areas not occupied by great reed warblers *(A. arundinaceus)*. In the presence of great reed warblers *(A. arundinaceus)*, some of the reed warblers *(A. scirpaceus)* nest in a less suitable habitat which is suboptimum, that is, in shrubs adjoining reeds (Jędraszko-Dąbrowska, unpublished).

These examples and many other results of removal experiments (e.g., Hoffmeyer, 1973; Redfield et al., 1977, Sheppe, 1977; Kozakiewicz, 1987) provide evidence that interspecific interactions, under certain circumstances, can limit movements or settlement of some species in some parts of the habitat. One species therefore can be a barrier to another species.

CONCLUSIONS

With respect to the scale of their effects, ecological barriers can be classified as acting at the level of individuals (barriers within a population area); at the level of populations or communities (barriers within metapopulation boundaries); and at the level of metapopulations (Figure 3).

Barriers at the level of individuals can be crossed by animals but they can limit the frequency of movements. This modifies the territorial behavior (the shape and size of home ranges or territories) of individuals occupying the area near the barrier. These barriers do not disrupt the population, but they can separate population parts with slightly different characteristics if the barrier occurs in a heterogenous habitat. Environmental barriers acting at the population level (within metapopulation boundaries) fragment the population. Such barriers strongly limit the movements of individuals or species, and may have a selective effect for individuals or species that can cross them. Parts of the population separated by barriers (subpopulations) form a metapopulation. Although individual subpopulations can show ecological differences, the metapopulation should be considered as an ecological population because its parts (subpopulations) can interact. These interactions, realized because of the movements of individuals across the barriers, can have an effect on both the ecological structure and the dynamics of individual subpopulations.

Environmental barriers limiting migrations (dispersal) of species have an effect on the structure and other characteristics of animal communities living in heterogeneous habitats. Ecological barriers are likely to affect animal communities in a similar way as they do populations. In any case, they influence the structure of competitive systems occurring in distinct patches of a heterogenous habitat.

REFERENCES

Banach, A. (1988). Bank vole population in a complex of forest biotopes. *Acta Theriol.*, 33:87–102.
Bakowski, C. and Kozakiewicz, M. (1988). The effect of forest road on bank vole and yellow-necked mouse populations. *Acta Theriol.*, 33:345–353.
Hansson, L. (1987). Dispersal routes of small mammals at an abandoned field in central Sweden. *Holarctic Ecol.*, 10:154–159.
Hoffmeyer, J. S. (1973). Interaction and habitat selection in the mice *Apodemus flavicollis* and *Apodemus sylvaticus*. *Oikos*. 24:108–116.
Kozakiewicz, A. (1987). Spatial distribution and interspecific interactions in a small rodent community of a lake coastal zone. *Acta Theriol.*, 32:433–447.
Kozakiewicz, M. and Jurasińska, E. (1989). The role of habitat barriers in woodlot recolonization by small mammals. *Holarctic Ecol.*, 12:106–111.
Kozakiewicz, M., Kozakiewicz, A. and Banach, A. (1987). Effect of environmental conditions on the character of spatial interactions among three small rodent species. *Bull. Acad. Pol. Sci. Cl. II*, 35:181–188.
Kozakiewicz, M. and Szacki, J. (1987). Small mammals of isolated habitats — islands on the mainland or only island populations? *Wiad. Ekol.*, 33:31–45. (In Polish with English summary.)
Levins, R. (1970). Extinction. *Some Mathematical Questions in the Life Sciences*, Vol. 2, M. Gerstenhaber (Ed.), American Mathematical Society, Providence, RI, 77–107.
Mader, H. J. (1984). Animal habitat isolation by roads and agricultural fields. *Biol. Conserv.*, 29:81–96.
Merriam, G., Kozakiewicz, M., Tsuchiya, E., and Hawley, K. (1989). Barriers as boundaries for metapopulations and demes of *Peromyscus leucopus* in farmland. *Landscape Ecol.*, 2:227–235.
Middleton, J. and Merriam, G. (1981). Woodland mice in a farmland mosaic. *J. Appl. Ecol.*, 18:703–710.
Oxley, D. J., Fenton, M. B. and Carmody, G. R. (1974). The effects of roads on populations of small mammals. *J. Appl. Ecol.*, 11:51–59.
Redfield, J. A., Krebs, C. J., and Taitt, M. J. (1977). Competition between *Peromyscus maniculatus* and *Microtus townscendi* in grasslands of coastal British Columbia. *J. Anim. Ecol.*, 46:607–616.
Rybacki, L. (1980). The Analysis of the Bird Community and Nesting Territories of Selected Species in a Coastal Zone of Sniardwy Lake. M.Sc. thesis, Faculty of Biology, Warsaw University, 38. (in Polish with English summary).
Sheppe, W. (1967). Habitat restriction by competitive exclusion in the mice *Peromyscus* and *Mus*. *Can. Field. Nat.*, 81:81–98.

7 Above-Ground Insect Biomass in Agricultural Landscapes of Europe

LECH RYSZKOWSKI
JERZY KARG
GRIGORE MARGARIT
MAURIZIO G. PAOLETTI
ROMAN ZLOTIN

Abstract: *Estimates of the above-ground biomass of insects were carried out in agroecosystems situated in two landscape types, mosaic and uniform, located in Poland and Romania. Further studies were carried out in mosaic landscapes located in Italy and in Russia.*

There were higher biomass figures with greater species diversities in the above-ground insect populations in the agroecosystems of mosaic landscapes as opposed to those which were uniform. These contrasts were caused by a higher abundance of refuges for insects in mosaic landscapes. A comparison of above-ground insects biomass from various regions of Europe indicated significant differences in the structure of insect communities, reflected in domination of large-body sized species from the order of Orthoptera in the southern regions (Romania and Italy) and small Diptera in those from the north (Poland and Russia).

INTRODUCTION

The simplification of the structure of agricultural landscapes due to intensive farming exerts a major impact on the standing crop and richness of the fauna of ecosystems (Ryszkowski, 1981, 1985). The comparison of estimates of the biomass of the total soil invertebrate community (excluding Protozoa) of mixed forests, meadows, and arable fields in moderate climate indicates their considerable impoverishment in agroecosystems. The biomass of animals living in the soil of arable fields is 12 to 30 times smaller than that of mixed forests, while that of meadows is 6 to 15 times smaller (Ryszkowski, 1985). Berger (1988) has documented a decrease in the standing crop of reptile and amphibian populations because of intensive farming. Economically significant pests, e.g., *Melolontha melolontha* show similar patterns, (Ryszkowski, 1981). A mosaic landscape, having elements such as shelterbelts, meadows, pastures, and small patches of water bodies supports the maintenance of a rich fauna. For instance, it was found that in mosaic landscapes the rich community of Apidae remained unchanged for many decades (Banaszak, 1985). A much richer avifauna has also been encountered (Gromadzki, 1970) with a higher abundance of some groups of insects (Karg et al., 1985; Ryszkowski and Karg, 1986; Karg, 1989; Paoletti et al., 1989).

This research enabled a hypothesis to be developed which suggested that the impoverishment of fauna due to intensive farming could be overcome by altering the mosaic structure of agricultural landscape through the introduction of a network of shelterbelts, stretches of meadows, small ponds, and other refuges.

In order to test this hypothesis, research was initiated on the biomass of aboveground insects in the mosaic and uniform types of agricultural landscapes in Poland and Romania.

In addition, studies compatible to the Polish mosaic agricultural landscape were carried out in the forest-steppe region near Kursk (Russia), and in a mosaic landscape in the vicinity of Padova (Italy). The aim of these additional studies was to evaluate the stress exerted by agriculture on insect communities by a comparison of their abundance in cultivated fields as opposed to grasslands having soils not disturbed by the activity of the farmers.

TERRAIN AND METHODS

Compatible studies were carried out in the contrasting types of agricultural landscapes in Poland and Romania during the years 1984 to 1988. In 1990 an evaluation of the above-ground insect biomass in mosaic landscapes was made in Italy and in Russia. In Poland and Romania fields cultivated with wheat, barley, maize, sugar beet, alfalfa, and grasslands were sampled. In both countries the studies were carried out in parallel, in both mosaic and uniform landscapes. The mosaic landscapes had many nonarable habitats, e.g., shelterbelts, hedges, meadows, small forests, and small ponds (western Poland), or strips of afforestation along a network of channels, and reed areas (Romania). Those shelter habitats occupied about 12% of the studied area in Poland and Romania. Both in Poland and Romania the sizes of fields in the mosaic landscape were small (from 2 to 30 ha), having a considerable crop diversity and varied crop rotation patterns. The uniform landscapes in both countries had no nonarable habitats. The sizes of fields exceeded 100 ha, with a very simplified crop rotation pattern, showing in many instances continuous cropping of the same cultivar. The research in the uniform landscapes was localized in the western part of Poland near Poznań and in southern Romania near Craiova.

In Italy the evaluation of above-ground insects was carried out in the vicinity of Pegolotte di Cona, Venezia (in May 1990). The mosaic landscape with wheat, potato, and alfalfa crops was rich in field shelterbelts. Grass ecosystems were studied in the seaside landscape of the Adriatic seacoast, Porto Tolle.

In Russia the evaluation was carried out in June 1990, in the vicinity of Kursk in a mosaic landscape containing fragments of natural steppes and small woodlands, with cultivation of wheat, sugar beet, and alfalfa, as well as in the natural, mowed, and grazed steppe.

All these comparisons were made using the quick-trap method (Ryszkowski and Karg, 1977) In the long-term investigations carried out in Poland and Romania samples were taken, on the average, every three weeks during the growing season (March through October), with ten samples in a series. A total of 5000 samples were collected, and the material included about 73,000 insects. In the supplementary evaluation made in Italy and Russia, the series included 5 samples (65 samples in Italy and 30 in Russia) which contained 1300 and 600 insects, respectively.

The results are presented in terms of biomass since this parameter better reflects the ecological functions fulfilled by the organisms than do estimates of density. The biomass estimation is of particular importance in analyzing predator-prey relationships, especially when the discrepancy between the body masses of each insect group is considerable. For example, one large insect can eat as much as several hundred small ones or a few medium-sized ones. Thus, in order to omit ambiguity in determination of the daily food intake, the analysis should be carried on in terms of biomass. In the analyzed material the mass of individual specimens ranged from about 0.005 mg dry weight (d.w.) (some Diptera and Hymenoptera) to several hundreds of milligrams d.w. (some beetles and Orthoptera). It was therefore necessary to present the results as biomass.

The trophic classification was made assuming dominating feeding habits of studied groups. *Aphididae and Thysanoptera* were excluded from the analysis because of the possible error in evaluating their density (Ryszkowski and Karg, 1977). The biomass was estimated on the basis of the mean mass of a given specimen for each taxon under consideration, multiplied by density estimates. The collected material was dry-preserved and determined with an accuracy up to the family or generic level.

RESULTS

Poland and Romania

The mean total biomass of above-ground insects in various crops from uniform landscapes of Poland ranged from 17.6 mg·d.w.·m^{-2} (maize) to 59.7 mg·d.w.·m^{-2} (alfalfa) from uniform landscapes, and 30.8 mg·d.w.·m^{-2} (maize) to 80.6 mg·d.w.·m^{-2} (alfalfa) from mosaic landscapes. On average, the insect biomasses in cultivated fields located in mosaic landscapes were higher by 34% than those living in fields of uniform landscapes (Table 1). Opposite results were obtained in small stretches of grasslands. Higher biomasses were found in small grasslands situated in uniform landscapes (60.9 mg·d.w.·m^{-2}) than in mosaic ones (45.2 mg·d.w.·m^{-2}) (Table 2). The grasslands were not plowed for several years and thus they were providing good shelter and refuge habitats for the development of insect larvae in the soil. In grasslands located in uniform landscapes, more

Table 1 Mean Biomass of Dominating Orders and Trophic Groups of Epigeic Insects in Uniform and Mosaic Landscape in Poland (mg · d.w. · m^{-2}).

Group	Uniform landscape						Mosaic landscape					
	Wheat	Barley	Maize	Sugar beet	Alfalfa	Mean	Wheat	Barley	Maize	Sugar beet	Alfalfa	Mean
Diptera	14.78	14.09	9.10	18.87	30.13	17.4	20.42	28.34	13.97	25.24	40.96	25.8
Coleoptera	5.81	1.78	4.79	6.34	11.89	6.1	3.84	7.42	8.95	10.44	15.04	9.1
Hymenoptera	1.10	0.50	0.56	1.58	4.50	1.6	1.46	1.17	2.16	2.78	5.71	2.7
Heteroptera	1.57	0.62	2.14	3.25	5.16	2.5	2.10	1.00	3.52	6.35	10.07	4.6
Homoptera	2.52	1.93	0.65	5.08	4.43	2.9	3.50	3.72	1.71	4.69	5.34	3.8
Orthoptera	0.00	0.00	0.00	0.00	1.73	0.3	0.35	0.00	0.00	0.43	1.73	0.5
Lepidoptera	0.02	0.02	0.00	0.02	1.59	0.3	0.05	2.16	0.01	0.63	1.22	0.8
Others	0.04	0.06	0.34	0.33	0.30	0.2	0.13	0.20	0.45	0.52	0.53	0.4
Saprovores	6.38	6.34	3.76	6.97	15.85	7.9	8.04	11.75	4.71	12.36	19.39	11.3
Herbivores	10.18	10.49	7.04	18.63	33.12	15.8	15.63	19.65	12.27	22.64	44.19	22.9
Predators	8.17	1.81	6.24	8.28	7.46	6.4	7.06	11.14	12.32	13.29	12.24	11.2
Parasitoids	1.11	0.35	0.54	1.59	3.30	1.4	1.12	1.47	1.47	2.79	4.73	2.3
Σ	25.84	18.99	17.58	35.47	59.73	31.5	31.85	44.01	30.77	51.08	80.60	47.7

Table 2 Mean Biomass of Dominating Orders and Trophic Groups of Insects in Grassland Habitats in Uniform and Mosaic Landscapes in Poland and Romania (mg · d.w. · m^{-2})

Group	Poland		Romania	
	Uniform landscape	Mosaic landscape	Uniform landscape	Mosaic landscape
Diptera	25.53	21.99	19.18	13.44
Coleoptera	3.58	5.02	155.98	16.06
Hymenoptera	2.83	3.55	12.31	13.99
Heteroptera	7.31	5.36	33.01	8.07
Homoptera	7.28	5.75	5.86	6.41
Orthoptera	13.91	0.77	340.78	214.91
Lepidoptera	0.37	2.19	16.10	6.41
Other	0.05	0.53	8.66	0.00
Total	60.86	45.16	591.88	279.29
Saprovores	6.43	9.78	113.92	3.06
Herbivores	41.76	24.20	448.57	256.08
Predators	11.46	9.35	26.08	16.72
Parasitoids	1.21	1.83	3.31	3.43

insects were attracted from larger areas than those in the mosaics, which accounts for the higher biomass detected. For instance, insects which fly over long distances (e.g., Shaeroceridae) had a much greater (up to four times) biomass in the uniform landscape than in the mosaic. In both landscape types in Poland, the dominating group, amounting to about 54% of the total biomass, was made up by Diptera; the Coleoptera biomass constituted only about 20% (Table 1).

In Romania the above-ground biomass of insects from crops cultivated in the uniform landscapes ranged from 73.1 mg·d.w.·m^{-2} (alfalfa) to 375.9 mg·d.w.·m^{-2} (barley) as compared to 100.8 mg·d.w.·m^{-2} (sugar beet) to 584.0 mg·d.w.·m^{-2} (barley) in mosaic landscapes. On average, the insect biomass is 36% higher in the mosaic landscape than in the uniform one (Table 3). Significant differences were found in the levels of biomasses between Poland and Romania. The biomass in Romania was about 79% higher, both in the uniform and in the mosaic landscapes. Comparison of biomasses of abundant insect orders showed that for the 40 available biomass comparisons (from 8 abundant systematic groups found in 5 examined crops) in Poland, higher biomass figures were found in 33 instances in the mosaic landscapes, but in Romania there were only 23 such cases. Only the specialized herbivorous (crop pests) species showed a greater biomass in the uniform landscapes. For example, species from the Iassidae and Chrysomelidae (*Lema* sp.) families, which frequently occur on cereal crops as pests, are responsible for the higher standing crops of Homoptera and Coleoptera in the uniform landscapes. The *Eurygaster integriceps* (Heteroptera) a pest species in Romania but which does not occur in Poland, can increase to over 20% of the total insect biomass in cereal crops situated in uniform landscapes, whereas in crops located in mosaic landscapes its biomass contribution ranges from 0 to 9% of the total biomass.

Table 3 Mean Biomass of Dominating Orders and Trophic Groups of Epigeic Insects in Uniform and Mosaic Landscapes in Romania (mg · d.w. · m^{-2})

Group	Uniform landscape					Mosaic landscape				
	Wheat	Barley	Maize	Sugar beet	Alfalfa	Mean	Wheat	Barley	Maize	Sugar beet	Alfalfa	Mean
Diptera	7.44	14.72	7.95	15.96	20.39	13.3	16.91	11.73	6.35	25.08	36.48	19.3
Coleoptera	26.18	13.47	17.45	16.58	7.76	16.3	9.16	12.96	27.79	24.01	21.97	19.2
Hymenoptera	4.32	4.67	5.41	6.11	7.28	5.6	3.51	4.80	13.00	7.86	9.02	8.6
Heteroptera	13.64	29.34	4.88	26.77	21.40	19.2	31.23	74.30	50.11	5.16	15.75	35.3
Homoptera	7.69	7.66	1.81	8.36	3.45	5.8	3.14	2.86	1.36	9.68	3.28	4.1
Orthoptera	34.38	304.10	45.57	32.26	6.96	84.2	36.65	468.60	68.57	22.03	110.20	141.2
Lepidoptera	0.94	0.00	1.71	5.23	5.63	2.7	0.00	3.25	6.95	6.14	0.63	3.4
Other	4.17	1.98	0.56	0.22	0.20	1.4	0.59	0.55	0.35	0.86	1.90	0.8
Saprovores	21.30	7.16	3.65	1.56	2.57	7.3	3.25	7.63	3.25	1.57	6.71	4.5
Herbivores	67.05	352.60	62.71	98.16	56.71	127.4	84.13	567.50	142.10	83.22	161.6	207.7
Predators	8.03	13.83	15.93	7.11	7.32	10.4	11.79	6.97	28.18	10.52	27.21	16.9
Parasitoids	2.38	2.38	1.05	4.66	6.47	3.4	2.04	1.92	0.96	5.51	3.67	2.8
Σ	98.76	375.90	83.34	111.50	73.07	148.5	101.20	584.00	174.5	100.8	199.20	231.9

Table 4 Mean Body Weight (mg · d.w./ind.) of Above-Ground Insects in Uniform and Mosaic Landscapes in Poland and Romania

Group	Poland		Romania	
	Uniform landscape	Mosaic landscape	Uniform landscape	Mosaic landscape
Diptera	0.59	0.71	0.68	0.68
Coleoptera	2.02	2.93	4.10	3.37
Hymenoptera	0.46	0.55	0.88	1.25
Heteroptera	1.78	1.77	6.62	8.31
Mean (all insects)	0.73	0.87	3.37	5.16
Saprovores	0.36	0.64	1.02	0.40
Herbivores	0.99	0.91	4.56	8.02
Predators	2.01	2.69	3.44	3.72
Parasitoids	0.45	0.51	0.56	0.44

In the grassland refuge habitats in Poland and Romania, a higher above-ground biomass of insects, reaching 53%, was observed in uniform landscapes (Table 2). The mean body weight in most of the insect groups was found to be higher in mosaic landscapes than in the uniform sites (Table 4). The above results were obtained from both Poland and Romania. The results therefore indicate better conditions for the development of larger species with longer development in the mosaic landscapes. A landscape having many refuge habitats favors development of large-sized insect species. In Poland these insects are mainly saprovores, predators (Coccinellidae), or even parasites (large-sized species from the Ichneumonidae family); whereas in Romania they are mostly herbivores.

In all the agricultural habitats studied in both Poland and Romania, the herbivore group predominates in the estimates of mean standing crop. On average, the herbivore domination reaches 50% in Poland and 80% in Romania (Tables 1 and 3). In most cases, all the trophic groups have a higher biomass in the mosaic rather than in the uniform landscapes. Among 20 of the cases analyzed (4 trophic groups studied in 5 habitats) in Poland, higher biomass figures were observed in mosaic landscapes (19 cases), whereas in Romania the same phenomena was observed in only 12 cases. In the case of predators, highly statistically significant differences in their standing crops were found between the two types of landscape. For instance, in Poland the predators' and parasites' biomass is some 70% higher in mosaic rather than uniform landscapes (Table 1).

In Polish conditions, a highly significant positive correlation was found between the biomass of predators and parasites with that of their potential prey (herbivores and saprovores). This correlation occurred in all the studied crops and in both landscape types, but in the uniform landscape the correlation coefficients were lower ($r = 0.52$) than in the mosaic one ($r = 0.71$). The value of correlation coefficients varied in different crops in the uniform landscapes of Poland from $r = 0.54$ (alfalfa) to $r = 0.77$ (barley), while in mosaic landscape

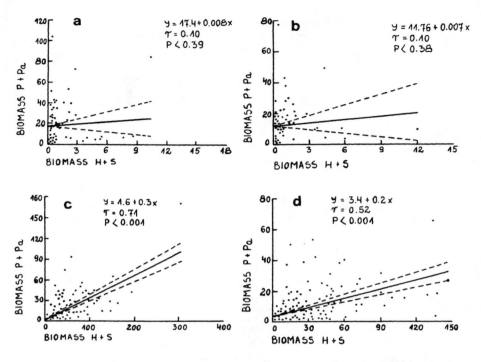

Figure 1. Correlation between biomass of predators (P) plus parasitoids (Pa), and herbivores (H) plus saprovores (S) in mosaic and uniform landscape in Romania and Poland in mg·d.w.·m^{-2}. P: biomass of predators, Pa: biomass of parasitoids, H: biomass of herbivores, and S: biomass of saprovores. (a) Romania, mosaic landscape; (b) Romania, uniform landscape; (c) Poland, mosaic landscape; and (d) Poland, uniform landscape.

the values ranged from r = 0.60 (barley) to r = 0.92 (maize). In Romania, however, a significant correlation was observed only in alfalfa fields, while in the other habitats analyzed no correlation was found (Figure 1).

Italy

The evaluation of insect biomass in three agroecosystems with mosaic landscapes in the vicinity of Padova are higher than those obtained in Poland in comparable situations. Orthoptera are the insects which cause such a high insect biomass level in Italy (particularly in cereal crops and alfalfa) (Table 5). The highest average biomass figures were observed in alfalfa (110.9 mg·d.w.·m^{-2}), and the lowest in potatoes (28.9 mg·d.w.·m^{-2}) (Table 5). In the agroecosystems studied in Italy, the contribution of Orthoptera to the total estimated biomass can exceed 55% (alfalfa). In grassland habitats, the large insects of the Orthoptera order make up 37% of the mean biomass estimate (Table 5). The mean biomass for grassland habitats was 92.3 mg·d.w.·m^{-2} and was similar to values observed in wheat and alfalfa.

Table 5 Mean Biomass (mg·d.w.·m^{-2}) of Dominating Orders and Trophic Groups in Mosaic Landscape and Grasslands in Italy

Group	Potato	Wheat	Alfalfa	Mean acroecosystems	Grassland
Diptera	9.33	25.35	34.45	23.24	19.79
Coleoptera	17.92	5.51	2.97	8.80	9.02
Hymenoptera	0.34	11.66	10.02	7.34	21.25
Heteroptera	0.00	1.36	1.00	0.79	4.74
Homoptera	0.60	0.54	1.56	0.90	1.78
Orthoptera	0.00	53.81	60.92	38.24	34.39
Lepidoptera	0.16	0.00	0.00	0.06	1.35
Other	0.00	0.33	0.00	0.11	0.00
Total	28.95	98.56	110.92	79.48	92.32
Saprovores	0.91	18.35	20.08	13.11	7.44
Herbivores	21.07	70.25	78.58	56.63	65.50
Predators	6.63	7.00	5.59	6.42	16.02
Parasitoids	0.34	2.96	6.67	3.32	3.36

Table 6 Mean Biomass of Dominating Orders and Trophic Groups of Above-Ground Insects in Russia (mg · d.w.·m^{-2})

Group	Sugar beet	Wheat	Alfalfa	Mean	Natural steppe	Moved steppe	Grazed steppe	Mean
Diptera	0.53	69.94	27.39	32.62	29.44	51.86	32.77	38.02
Coleoptera	2.97	14.33	0.44	5.91	15.78	16.18	1.74	11.23
Hymenoptera	0.00	2.13	6.18	2.77	6.15	6.28	5.22	5.88
Heteroptera	0.00	0.00	67.77	22.58	0.00	10.71	1.73	4.15
Homoptera	0.00	3.79	19.99	7.93	21.74	8.95	80.02	36.90
Orthoptera	0.00	0.00	0.00	0.00	32.49	32.49	44.66	36.55
Lepidoptera	0.75	0.00	0.00	0.25	1.51	46.05	1.51	16.36
Other	0.00	0.00	0.07	0.02	0.00	0.00	0.00	0.00
Total	4.25	90.19	121.84	72.08	107.11	172.52	167.65	149.09
Saprovores	0.00	6.47	7.87	4.78	13.02	17.12	12.95	14.36
Herbivores	1.85	53.30	107.26	54.15	76.06	126.63	145.05	115.91
Predators	2.40	29.89	1.03	11.08	13.99	21.26	1.70	12.32
Parasitoids	0.00	0.53	5.68	2.07	4.04	7.51	7.95	6.50

Russia

The results obtained for alfalfa and wheat are close to those recorded in Italy. The mean biomass of the above-ground insects in wheat was 90.2 mg·d.w.·m^{-2}, and on alfalfa 121.8 mg·d.w.·m^{-2} (Table 6). The dominating insect groups were Diptera (wheat) and Heteroptera (alfalfa).

In steppe habitats high mean biomass values were found, with the highest being in mowed steppes (172.5 mg·d.w.·m^{-2}) and the lowest in a protected natural steppe (107.1 mg·d.w.·m^{-2}) (Table 6). In the habitats of both the mowed and natural steppes large-body species predominated — Diptera, Orthoptera, and Lepidoptera. Whereas in the grazed steppes the species are small and are from the families of Iassidae and Delphacidae (Homoptera). The mean body

Table 7 Mean Biomass of Above-Ground Insects in Agroecosystems and Grassland in Mosaic Landscapes of Different Regions of Europe (mg $\cdot$ d.w. $\cdot$ m^{-2})

Country	One-year crops	Alfalfa	Grasslands
Poland	39.4	80.6	45.2
Russia	47.2	121.8	149.1
Romania	240.1	199.2	279.3
Italy	63.7	110.9	92.3
Mean	97.6	128.1	141.5

mass of individual insects in the natural steppe and in the mowed steppe habitats was 1.49 and 1.57 mg d.w., respectively, while in the grazed steppe this value amounted to only 1.18 mg d.w. The trophic structure of the above-ground insects also differed between the grazed steppe and the mowed or natural situations. The contribution of predators and parasites to the mean biomass of the grazed steppe is considerably lower than in the other steppe types (Table 6). In the grazed steppe predators and parasites make up 5.7% of the total mean biomass, in the mowed steppe, 16.6%, and in the natural one, 16.9%.

Comparison of the Biomass of the Above-Ground Insects in Different Regions of Europe

Supplementary studies carried out in 1990 in the vicinity of Padova and Kursk facilitated the initial comparison of insect biomass both in agroecosystems and grassland habitats of various European geographical regions (Poland, Romania, Italy, and Russia).

In the mosaic landscapes of the four studied countries, the highest mean biomass of above-ground insect ecosystems was observed in Romania whereas the lowest was recorded in Poland. The above-ground insect biomass in southern regions of Europe (Italy and Romania) on the average is twice as high as those in the northern ones (Poland and Russia) (Table 7).

In all the cases studied, a significant impoverishment of the above-ground insect communities was observed in cultivated fields as opposed to alfalfa and grasslands. This phenomenon is characterized by lower insect biomass figures in the northern habitats and by an appreciable simplification of their taxonomic and trophic structures. The exception was one site in Romania, where high mean values of the above-ground insect biomass were found in agroecosystems with a spring cereal cultivation which showed figures of biomass similar to those in grasslands biomass (Table 7). The high mean biomass estimates in this cultivation resulted from the large outbreaks of crop pests (e.g., Eurygaster and Orthoptera). In general, the taxonomic and trophic structure of the above-ground insect community in Romania agroecosystems had been significantly simplified. In Romania herbivores dominated in most cases, whereas parasites and predators had the lowest representation (Table 8). The taxonomic structure of insect commu-

Table 8 Contribution of Taxonomic and Trophic Groups in Acroecosystems of Different Regions of Europe (%)

Group	Poland	Russia	Romania	Italy
Diptera	54.1	45.2	8.3	29.2
Coleoptera	19.1	8.2	8.3	11.1
Hymenoptera	5.7	3.8	3.7	9.2
Heteroptera	9.6	31.3	15.2	1.0
Homoptera	8.0	11.0	1.8	1.2
Orthoptera	1.0	0.0	60.9	48.1
Other	2.5	0.5	1.8	0.2
Saprovores and herbivores	71.7	81.7	91.5	87.7
Predators and parasitoids	28.3	18.3	8.5	12.3

nities in the southern regions of Europe is characterized by the domination of large-sized Orthoptera species, which make up 60% of total mean biomass in Romania and 48% in Italy. In the northern regions of Europe the Diptera dominate and make up 54% of the total biomass of the insect community in Poland, and 45% in Russia (Table 8).

CONCLUSION

Long-term studies on the agroecosystem insects carried out in parallel in uniform and mosaic agricultural landscapes showed that the mosaics in landscapes play an important role in maintaining the richness of insect fauna. This function results from the existence of numerous refuges for insects and for potential dispersion of insects from patches of suitable habitats in the mosaic landscape which are not too remote. The result is that there is a high diversity and abundance of insects in the mosaic landscapes. The predatory and parasitic segment of insect communities, the occurrence of which depends considerably on the existence of refuge habitats, is particularly influenced by changes in landscape structure. Herbivores which are less limited by refuges than the availability of food, are less influenced by simplification of the landscape structure, and can thus maintain relatively high abundances despite changes in the pattern of landscape structure. Insect communities in the Romania agroecosystems have the most simplified taxonomic structure, with herbivores predominating. The numerical relations between the predators and their prey, under Romanian conditions, are so weak that a statistically significant correlation between the biomass of these insect groups was found only in the case of alfalfa. In Poland, the majority of the crops under investigation showed a highly significant correlation between the biomass of predators and herbivores or saprovores.

The studies carried out in grassland ecosystems provided very interesting results. Both in Poland and Romania the grassland habitats located in uniform landscapes are characterized by a higher insect biomass than those in mosaic landscapes. It seems that grasslands surrounded by large homogeneous crop fields provide good shelter and refuge places for many insects, and because of this function they are places in which insects are concentrated.

The supplementary studies carried out in Italy and in Russia made it possible to make wider comparisons of the above-ground insect biomass in agroecosystems and grassland habitats. In all cases (Poland, Romania, Russia, and Italy) an impoverishment of the insect fauna was found in the cultivated fields.

The evaluations carried out in the steppe environment (in the vicinity of Kursk) proved that the human pressure (mowing and grazing) brings about an increase in the total above-ground insect biomass as compared to the natural steppe, it also induces changes in their taxonomic and trophic structure. The negative effects of grazing by cattle have been especially noted in the steppe ecosystems, and involve a substantial increase in the number of small-sized herbivorous species (Homoptera) and a corresponding decrease in the number of predators and parasitoids.

REFERENCES

Banaszak, J. (1983). Ecology of bees *(Apoidea)* of agricultural landscape. *Pol. Ecol. Stud.*, (9)4:421–505.

Berger, L. (1988). An all-hybrid water frog population persisting in agrocenoses of central Poland. *Proc. Natl. Acad. Sci. U.S.A.*, 40:163–173.

Gromadzki, M. (1970). Breeding communities of birds in mid-field afforested areas. *Ecol. Pol.*, 18:307–350.

Karg, J. (1989). Zróżnicowanie liczebności i biomasy owadów latających krajobrazu rolniczego zachodniej Wielkopolski. *Roczniki Akad. Roln. w Poznaniu*, 188:5–78.

Karg, J., Margarit, G., Hondru, N. and Gogoasa, C. (1985). Preliminary data regarding the influence of landscape types on the epigeic insects from wheat and alfalfa crops. *Prob. Protect. Plantetor (Bucharest)*, 3:289–302.

Paoletti, M., Favretto, M., Ragusa, S. and Strassen, R. (1989). Animal and plant interactions in the agroecosystems. The case of woodland reminants in northeastern Italy. *Ecol. Int. Bull.*, 17:79–91.

Ryszkowski, L. (1981). Wptyw intensyfikacji rolnictwa na faunę. *Zesz. Probl. Post. Nauk Rol.*, 233:7–38.

Ryszkowski, L. (1985). Impoverishment of soil fauna due to agriculture. *Soil Ecology and Management*, J. H. Cooley, (Ed.), *Intecol Bull.*, 12:6–17.

Ryszkowski, L. and Karg, J. (1977). Variability in biomass of epigeic insects in the agricultural landscape. *Ekol. Pol.*, (25)3:501–517.

Ryszkowski, L. and Karg, J. (1986). Impact of agricultural landscape structure on distribution of herbivores and predator biomass. Impact de la structure des paysages sur le protection des cultures. *Colloq. L'INRA*, 36:39–50.

8 Survival of Populations and the Scale of the Fragmented Agricultural Landscape

J. T. R. KALKHOVEN

Abstract: *In the fragmented landscape of Western Europe, populations in small habitat patches show large fluctuations in density or even local extinction. A species sensitive to fragmentation can only survive if dispersal and colonization occur at a sufficient rate. To prevent regional extinction the small populations should be connected in a network: a metapopulation.*

The effect of fragmentation can be diminished by (1) improving the quality and the size of the habitat patches, (2) increasing the number of patches, and (3) decreasing the resistance of the landscape. The measures should be taken at the scale on which the species uses the landscape.

INTRODUCTION

In the agricultural landscape of western Europe, populations of many species occur in relatively small landscape units. They often show large fluctuations in density or even local extinction and can only survive if dispersal and colonization occur at a sufficient rate. The system of interrelated populations of a species in a certain area is called a metapopulation and is typical of a fragmented landscape.

Further fragmentation of more or less natural habitats by agricultural use, traffic, and urbanization causes changes in habitat area, habitat quality and quantity, and increases distance to other likely habitats. The effect on species survival depends on the population density, the rate of dispersal, and the scale on which the species uses the landscape.

There are relations between small landscape elements on a local scale of hundreds of meters and between landscapes on a regional scale from about ten to a few hundred kilometers. The metapopulations can be studied by means of pattern and process analysis and modeling. For many species that are in danger of (regional) extinction, improvement of the habitat quality and the landscape structure is necessary.

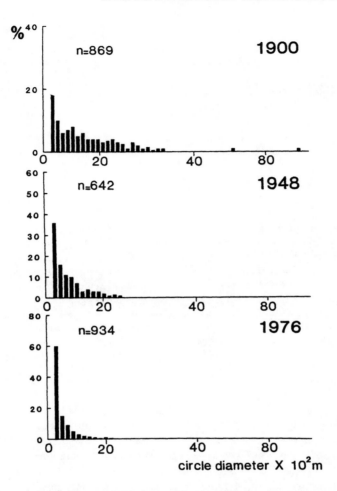

Figure 1. Size distribution of natural areas in The Netherlands. The surface of the nature areas is transformed into circles. The change in area within a century is mainly caused by fragmentation. (Source: Ministry of Agriculture, Nature Conservation and Fisheries, 1989.)

FRAGMENTATION

In most countries of the northwestern European lowlands, 60 to 70% of the area is used for intensive agriculture and about 10% is urban land. The remaining 20 to 30% consists of lakes, forest, and other areas of natural vegetation. Most of these more or less natural areas are small — often less than 100 ha, or even less than 10 ha (Figure 1). The modern agricultural landscape consists of a matrix of farmland and scattered small landscape elements such as forest patches, hedgerows, wooded banks, road verges, pools and ponds, and ditches and watercourses. Heathland, marshland, and hay meadows covered hundreds of hectares half a century ago, but are nowadays often restricted to small remnants. A large part of the western European landscape, especially on sandy soils, has long been a heterogeneous countryside with a mixture of different-sized patches of (semi)natural elements in a matrix of cropped land. This small-scale landscape, as it is called (De Regt, 1989), will vanish rapidly unless special conservation measures are taken. Hedgerows are being removed, fields enlarged, and wet meadows and marshlands drained and cultivated causing small elements of native vegetation to diminish in number and in area. The landscape has become more homogeneous, with larger agricultural fields and fewer and smaller natural areas. For many species, such a landscape is fragmented.

Many wild animals and plants depend on individual landscape elements or a special mosaic of elements. The red list of plants in The Netherlands now contains 541 species (Weeda et al., 1990); about one fourth of these species grow in the small landscape elements mentioned above. There are 12 of the 57 species on the national red list of threatened birds that are characteristic of small-scale, heterogeneous, agricultural landscapes (Osieck, 1986). About one third of the butterflies in the western European lowland are restricted to small and linear strips of land that are rich in flowering plants. These can especially be found on the edges of farmland and forest patches, and also along road verges and hedgerows. For several of these species, the progressive fragmentation represents a threat to survival.

Further fragmentation of more or less natural habitats by agricultural use, traffic, and urbanization cause changes in their area, their quality and quantity, and increases the distance to other similar habitat patches. The effect on species survival is discussed in this chapter.

HABITAT PATCH AND LOCAL EXTINCTION OF POPULATIONS

The landscape elements have several ecological functions. Firstly, they serve as habitat patches or part of habitat patches for wildlife and, secondly, as corridors or stepping stones. As a result of fragmentation, habitat patches are mostly small and separated from each other by more or less hostile areas: a matrix which is not suitable as a habitat.

Small habitat patches can only support small populations of plants and animals. How large a population can become also depends on the species concerned and on the quality of the patch as a habitat. Millipedes are expected to be numerous even in a small forest patch of 5 ha, if it is moist enough. In the same patch, only a few red squirrels can live. In a woodlot of the same size, but with a certain proportion of coniferous trees, a higher number can be expected compared to a deciduous woodlot; but even in an optimal habitat no more than two or three individuals per hectare will normally occur (Verboom and Van Apeldoorn, 1990).

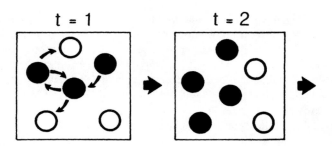

Figure 2. Spatial dynamics in a metapopulation. At two different points in time (t = 1 and t = 2) the patterns of occupied (black circles) and unoccupied habitat patches (open circles) are different. Two patches are still occupied, two are colonized, one remains empty and one has become empty.

The quality of a patch as a habitat can range between optimal and marginal. This range can be seen in space as well as in time. For example, hedgerows in different places will show at least slight differences in quality as a habitat for breeding birds. A small marshland can be an optimal habitat for some species in wet years, but a marginal one in dry years, for the same species. In these marginal situations there is a great chance that populations of these species are very small, or may greatly diminish.

Small populations have a great chance to degenerate genetically or even to become extinct locally, within a relatively short period of time. There is a growing amoung of literature showing local extinction of small populations being quite common in animals and short-lived plants (Opdam, 1988). Suitable habitats can become empty for a certain period of time caused simply by stochastic changes in the local environment or in the populations of the organism itself (Verboom et al., 1991). In extremely cold winters (not predictable) the number of wrens *(Troglodytes troglodytes)* may be decimated. An unexpected disease can have the same effect on several species. All these processes will have an effect on the occupancy of the suitable patches in the landscape (Figure 2).

LANDSCAPE STRUCTURE AND DISPERSAL

Local extinction can be prevented if there is some degree of exchange of individuals among the small populations and can be compensated when enough individuals can reach an empty habitat. This exchange is called dispersal. By means of dispersal weak populations can be supported and empty habitat patches can be (re)colonized (Opdam, 1990). The effect of dispersal movements is influenced by the distance between the habitat patches, by the resistance of the landscape in between, and by the behavior of the organism concerned. When the scale of the landscape is coarse grained and an animal can only disperse over short distances, several suitable but isolated habitat patches can hardly be reached or, perhaps, cannot at all be colonized by this animal. Several landscape elements can function as corridors, which enhance dispersal efficiency. Animals such as the badger *(Meles meles)*, the bank vole *(Clethrionomus glareolus)*, and some birds follow hedgerows while migrating through the landscape. Barriers such as busy roads and canals may be a hindrance for animals in their attempt to reach other habitat patches. Such obstacles enlarge the resistance of the landscape.

METAPOPULATION SURVIVAL

The set of (small) fragmented populations, connected with each other by means of dispersal, is called a *metapopulation* (Levins, 1970; Opdam, 1988). It is a system where the elements, the various subpopulations, are unstable, but which is stable on a higher level. Even if the elements are very small, survival of the whole system is possible. The survival of several animal species has been studied in the fragmented landscape of The Netherlands by analyzing the pattern of occupancy of habitat patches and the change over a few years. Additional analysis was carried out with the help of a computer model which enabled analysis over many generations, that under field conditions is not possible. The most important hypothesis and results will be given here.

The survival of the metapopulation depends upon the number and the size of subpopulations and the rate of dispersal. The more suitable patches there are at a bridgeable distance, the greater the chance for exchange and recolonization.

Usually, the patches vary in size, and consequently the maximal size of the subpopulations varies. Two extreme models can be distinguished (Figure 3). The first is the model of an archipelago system, where the patches are all small. This system is vulnerable to extinction when the local extinction of the subpopulations is not compensated by recolonization. This is the situation for the tree frog *(Hyla arborea)* in The Netherlands which has small populations with little or no exchange of individuals (Stumpel and Hanekamp, 1986). To permit dispersal at a sufficient rate, the distance between the habitat patches of the tree frog must be less than 1 km, preferably less than 500 m.

In the other model, the system consists of a large area with a large permanent population, and a series of small patches. In this system, there is always a stream (supply) of individuals from the source area to the small habitat patches. It is more stable, although there may be a constant change in occupation of the small patches. This model holds for the nuthatch *(Sitta europaea)*, a bird which lives in old deciduous forest patches (Verboom et al., 1991). Recolonization depends on the distance between the habitat patches, which must not be more than 2 km. The same holds true for the red squirrel and several forest birds (Verboom and Van Apeldoorn, 1990; Opdam, 1991). The chance for occupation of suitable forest patches is significantly higher in the neighborhood of a large forest, which can act as a permanent supply source (Van Dorp and Opdam, 1987). These authors suggested that, on a larger scale, the gradient of declining forest density

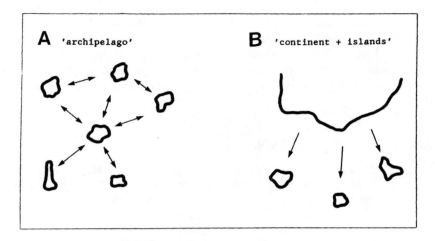

Figure 3. Two examples of relationships among habitat patches in a fragmented landscape. (A) All patches are about equal in size and may become empty due to local extinction; patches are interdependent. (B) A large patch serves as a continuous source of dispersers, supporting subpopulations in small patches and, hence, lowering local extinction chance here. Extinction rate in the large patch is assumed to be negligibly small (From Opdam [1990], *Species Dispersal in Agricultural Habitats,* R. G. H. Bunce and D. C. Howard (Eds.), Bellhaven Press, New York. With permission.)

from western Germany to central Holland is reflected in a decrease in average probability of occurrence.

With the help of a stochastic computer model the survival time for a metapopulation of the badger could be calculated (Lankester et al., 1991). Unlike deterministic models, this stochastic model accounts for accidental demographic events, which play an important role in small populations. In the Dutch landscape, badgers live in clans of up to five adult animals in a sett, generally situated in the forest. The most common cause of death is by collision with traffic, which occurs even on small roads. Exchange of animals between the subpopulations will prolong the survival time of the small groups. Dispersal is likely to be facilitated by a certain density of corridors, which can give cover and food. The survival of a badger metapopulation will also be enhanced by a larger number of connected setts (Figure 4).

The metapopulation concept seems to hold for some plant species (Van Ruremonde and Kalkhoven, 1991), but the slow turnover in long-lived species and the presence of a seed bank provide problems for further research.

RELATION WITH LANDSCAPE MANAGEMENT

The survival of many species in the fragmented landscape depends on processes at the level of the metapopulation. The most important processes are local ex-

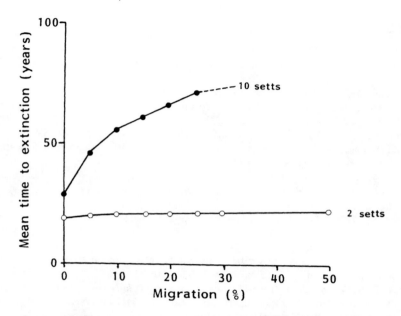

Figure 4. Mean time to extinction of a badger metapopulation. The survival chance in an area is dependent on the number of setts and the dispersal rate. (From Lankester, K., Van Apeldoorn, R., Meelis, E. and Verboom, J. [1991], *J. Appl. Ecol.,* . With permission.)

tinction, which is dependent upon patch size and quality and influenced by stochasticity in the environment, and (re)colonization, which is influenced by the distance between habitat patches and the presence of corridors and barriers. Dispersal is thus the key to survival of the metapopulation (Opdam, 1990).

Based upon this basic landscape ecological knowledge, the following rules for landscape management can be formulated, in order to increase the survival chance of threatened species in the fragmented landscape:

- Increase the size and the quality of the habitat patches in order to increase the local population size and to diminish risk of extinction;
- Increase the number of patches in order to improve the possiblity for exchange and recolonization, and to lower the change of stochastic extinction of the metapopulation; and
- Decrease the resistance of the landscape by including corridors and reducing the effect of the barriers, in order to enhance the possibility of dispersal.

On what scale the measures, mentioned in these rules, are to be taken will always be dictated by the specific species and by the way it lives and behaves in the landscape.

REFERENCES

De Regt, A. L. (1989). Small-Scale Landscape in a Large-Scale Europe. Ministry of Housing, Physical Planning and Environment, National Physical Planning Agency, The Hague, 36.

Lankester, K., Van Apeldoorn, R., Meelis, E. and Verboom, J. (1991). Management perspectives for the populations of the Eurasian Badger *(Meles meles)* in a fragmented landscape. *J. Appl. Ecol.*, 28:561–573.

Levins, R. (1970). Extinction. Some mathematical questions in biology. Lectures on Mathematics in Life Sciences, 2, M. Gerstenhaber (Ed.), American Mathematical Society, Providence, RI, 77–107.

Opdam, P. (1988). Populations in fragmented landscape. *Connectivity in Landscape Ecology*, K.-F. Schreiber (Ed.), 2nd Int. Semin. Int. Assoc. Landscape Ecol., Münster, 75–77.

Opdam, P. (1990). Dispersal in fragmented populations: the key to survival. *Species Dispersal in Agricultural Habitats*, R. G. H. Bunce and D. C. Howard (Eds.), Belhaven Press, New York, 3–17.

Opdam, P. (1991). Metapopulation theory and habitat fragmentation: a reivew of holarctic breeding bird studies. *Landscape Ecol.*, 5:93–106.

Osieck, E. R. (1986). Bedreigde en karakteristieke vogels in Nederland (Threatened and characteristic birds species in the Netherlands, with English summary). Nederlandse Vereniging tot Bescherming van Vogels, Zeist, 132.

Stumpel, A. H. P. and Hanekamp, G. (1986). Habitat and ecology of *Hyla arborea* in the Netherlands. *Studies in Herpetology*, Z. Rocek (Ed.), Charles University, Prague, 409–411.

Van Dorp, D. and Opdam, P. F. M. (1987). Effects of patch size, isolation and regional abundance on forest bird communities. *Landscape Ecol.*, 1:59–73.

Van Ruremonde, R. H. A. C. and Kalkhoven, J. T. R. (1991). Effects of woodlot isolation on the dispersion of fleshy-fruited species. *J. Vegetation Sci.*, (in press).

Verboom, B. and Van Apeldoorn, R. (1990). Effects of habitat fragmentation on the red squirrel, *Sciurus vulgaris* L., *Landscape Ecol.*, 4:171–176.

Verboom, J., Lankester, K. and Metz, J. A. J. (1991). Linking micro and macro levels in stochastic metapopulation dynamics. *Biol. J. Linn. Soc.*, 42:39–55.

Verboom, J., Schotman, A., Opdam, P. and Metz, J. (1991). European nuthatch populations in a fragmented landscape. *Oikos,* 61:149–156.

Weeda, E. J., Van der Meijden, R. and Bakker, P. A. (1990). FLORON-rode lijst 1990. (FLORON Red Data List 1990. Red data list of the extinct, endangered and vulnerable plants in the Netherlands in the period 1980–1990, with English abstract). *Gorteria,* 16:1–26.

9 The Implications of Scale on the Ecology and Management of Weeds

L. G. FIRBANK

Abstract: *Weed management must be seen as an ecological problem. Ecological systems display properties of hierarchy and scale, and therefore particular experimental approaches to a given problem can offer different perspectives. Spatial and temporal scales and levels of organization should be specified when studying the ecology and management of weeds, and the difficulties of extrapolating from one scale to another should be appreciated. Unfortunately, most weed research has taken place at smaller scales than required by farmers and planners, and even though larger scale studies are now becoming more common, they are not without practical and statistical problems. The lack of replication of some field-scale studies can be countered if there is greater transfer of information from farmers to each other and to researchers, rather than the more formal situation of information transferring solely from researchers to farmers.*

INTRODUCTION

There is increasing interest in the development of approaches to farming which can integrate biological, cultural, and often chemical methods of control. The idea of such integrated pest management has a long history now. Early classical biological control programs dealt with the restoration of predatory-prey balances which had been disrupted, either by the invasion of new pests in the absence of their predators or pathogens, or by damaging inputs of agrochemicals. More recently, it has become clear that these successes can be built upon by more elaborate approaches. It is possible to manage the farming landscape to promote some species and not others. Also, it is possible to forecast the dynamics of some pest species so as to allow the efficient, rather than indiscriminate, use of agrochemicals.

The application of such integrated approaches to agriculture are dependent upon a detailed understanding of the underlying ecology of at least some of the pests, diseases, and weeds. One needs to know how the populations of particular species will change in response to possible types of management by the farmer.

It is not the author's intention to review how this may be done to maximum effect. It is rather to explore the implications of scale effects on the study and analysis of the agroecosystem, and to show how scale effects can help us direct future research in integrated pest control.

Ecological Levels of Organization

Considerations of scale are central to ecology. Indeed, one of the signs that distinguished true ecology from the inquiring natural history of the nineteenth century was the development of ideas concerning levels of organization in ecological systems. Individual organisms form local, single-species populations. Populations of different species in the same area form communities, where species interactions can influence the abundance of individual populations. On a larger spatial scale, communities combine to form ecosystems, complexes of communities which nevertheless may be constrained by limitations from climate, nutrient, and energy flows. Finally, all ecosystems on the earth are linked through the sharing of the atmosphere (e.g., Begon et al., 1990).

Debates about the precise role of the levels of organization continue to this day. There are those (e.g., Clements in the early part of the present century and Lovelock in the latter part) who place emphasis on a holistic vision of nature, where the ecosystem and the biosphere take on some characteristics of living creatures — development and self regulation. In contrast, others such as Tansley, the father of British ecology, and the schools of Imperial College, Oxford and York during the 1970s, promote a much more reductionist view; that by understanding the behavior of the components you can understand the behavior at larger scales (see, e.g., Sheail, 1987, for a review of some of these debates). There are clearly philosophical aspects to these disagreements, but they should not be misrepresented as conflicts between hard-headed reductionists and woolly, sentimental holists. The scientific debate continues into the present (see, e.g., Lovelock, 1979; Carney, 1990 and included references).

Fundamental to some of these debates is confusion about the very nature of hierarchies of organization. Let us take a deliberately contentious analogy — the human body — corresponding to the ecosystem. The cells correspond to individuals, which form populations in tissues and 'communities' as organs. Information is passed up and down the hierarchy. Thus, an infection can be detected by monitoring the population size of lymphocytes in the blood stream, by observing swelling of the affected organ, and possibly of other organs (especially the lymph glands), and by monitoring the temperature of the patient. Different levels of investigation require different techniques, and are aimed at collecting different types of data. Moreover, if knowledge about the human body was restricted, it would be very difficult to forecast that simply taking the temperature might be a more efficient way of examining the health of the system than monitoring lymphocytes.

We know far less about ecosystems than about the human body, and the way in which different parts of the system interact is often poorly understood. However, ecosystems do display properties which are best studied at certain scales, even though they may emerge from the behavior at lower scales. For example, in monocultures of plants the development of plant buds depends upon the available of resources — light, water, and nutrients. Larger and faster growing plants can often access these resources more easily than smaller, shaded plants, and so can produce more buds than smaller neighbors, exaggerating discrepanices in plant size. However, the total amount of resources per unit area limits the total number of buds per unit area, and hence, the total yield per unit area. The limitation of total yield per unit area can be regarded as an emergent property of competition between neighboring plants (e.g., Firbank and Watkinson, 1985a). If one is interested in crop yield, it is more efficient to sample a number of areas which display the emergent property of yield restriction, than isolated plants which do not. On the other hand, if one is interested in the effects of competition on natural selection within the plant stand, one would have to look at variation between individual plants, as natural selection operates at the individual plant level.

The essence of an hierarchy is that each level consists of units that can be studied in isolation, as combinations of subcomponents, or as subcomponents in themselves of a yet higher level of organization. These units are referred to as holons (Koestler, 1967). Information is transferred from one level of the hierarchy to neighboring levels. The problem with a reductionist approach to science is that individual components are studied intensively, but the interactions between components and between levels of organization are understood much less well. In particular, care must be taken when extrapolating from reductionist studies to ensure that the properties considered by them really do affect higher levels of organization — that they really do contribute to emergent properties.

It is therefore most important that the problem under study is clearly defined, and that extrapolation from small-scale reductionist experiments to large-scale systems may be misleading.

SCALE EFFECTS AND WEED ECOLOGY

The Spatial Dimension

Weeds, like all plant species, show different responses at different levels of spatial pattern. The effects of intra- and interspecific competition are very local, and can be assessed by planting out the weeds over a range of densities and finding the density at which a decrease in mean plant size becomes apparent. The inverse of the density gives an estimate of the area required by a plant to grow to the maximum size within the given conditions, and hence estimates the range over which the effects of competition can be felt by the individual plant (Firbank and Watkinson, 1985a).

The procedure is to model the mean yield per plant, $\bar{w}$, of the weed in monoculture as a function of density N:

$$\bar{w} = w_m (1 + a N)^{-b} \qquad \text{(Equ A)}$$

where w_m is the mean yield of isolated plants and a is the parameter which estimates the area to grow to that size [n.b., the quality of the estimates can decline if the power term b is markedly different from 1 (Firbank and Watkinson, 1985a)].

For the grass weed *Bromus sterilis*, a value of 0.2 m² was obtained for a from plants grown in the field (Firbank et al., 1984) and for the dicotyledonous *Agrostemma githago* a value of 0.06 m² was given (Firbank and Watkinson, 1985b). It is difficult to generalize from two figures, but it seems that competitive effects can be noticed at distances of less than 1 m, although indirect effects can occur where a plant is reduced in size because of competition from a neighbor (Harper, 1977).

The next aspect is the spatial distribution of the plant within the field. Even a casual glance at a weedy field reveals that the weeds are not dispersed uniformly nor at random; typically they form clumps or strips. Sometimes the cause is obvious — ingress from the field edge (Marshall, 1989), or evidence of less-than-thorough herbicide application; sometimes it is less obvious, and may result from variation in soil type, or from dispersal from an isolated colonist in the past. Marshall (1988) used 0.25 m² quadrants at different sampling intensities to investigate this problem, and found that the dispersion of three grass weeds *B. sterilis, B. commutatis,* and *Elymus repens* were all highly skewed — some quadrats having many plants, and many quadrats having none. He suggested that at least 18 locations ha^{-1} are needed to obtain reasonable estimates of mean density.

Increasing the spatial scale, we come at last to a clearly human-induced level: the field. Repeated experiments show how management affects weed populations both in the short term and in the longer term (see, e.g., Firbank [1991] for a recent review). There is no doubt of the importance of field-scale management practices on weed populations, although it is worth remembering that this need not be true for all organisms in the acroecosystem. Aebischer (1991), for instance, records that in a 20-year survey on five farms, many invertebrates showed few differences between fields and farms, and seemed to be changing evenly across the whole area in response to larger-scale factors.

The degree of interaction between weeds from neighboring fields is unclear. The idea that weeds may constitute metapopulations, consisting of local subgroups which may become extinct, but which can be recolonized from other subgroups, is attractive, but the author knows of no formal analysis of this issue. Finally, at the largest spatial scale, we encounter restrictions of the range of weed species, whether due to climate or other factors.

These different scales can be regarded as forming an hierarchy (O'Neill, 1989). Different quadrat sizes will naturally detect different levels of pattern (e.g., Kershaw and Looney, 1985), and the choice of quadrat size should be influenced by the objective of the study. Yet, in practice this has rarely been the case; the choice of quadrat size, or experimental unit, has depended upon practical constraints which are related only weakly to the true objectives. Our own experiments on weed-crop competition are typical: in the past we have used plot sizes of between 1 and 9 m² and sown the weeds at random within the plots; in this way it is possible to get very clear relationships between crop yield and weed density (e.g., Firbank et al., 1984; Firbank and Watkinson, 1985b). The experiments were designed to produce estimates of error about the mean yield for a given density, either by using replication or a wide range of densities (Cousens et al., 1988). Unfortunately, the statistical population being sampled by such an experiment is restricted to one part of one field of randomly dispersed plants in one season. If the true statistical population under study is something more practical — cereal fields within a small geographic region, for example, these error estimates have no value — we would have mistaken subsampling for replication (Eberhardt and Thomas, 1991). The whole trial needs to be repeated in different places and at different times if genuine replication is to be achieved.

This has certainly happened, but not on many occasions. The general relationship between crop yield and weed density can usually be described by the same kind of hyperbolic model (Cousens, 1985), but while Cousens et al. (1988; and Firbank et al., 1990) report significant differences between the model parameters from place to place and from soil type to soil type at the same place, Gill et al. (1987) note more consistency among infestations of *Bromus diandrus* in southwestern Australia. The Streibig et al. (1989) analysis of Australian weed infestations suggests that while there are differences between sites and years for some species, the economic control thresholds tend to be within an order of magnitude of each other for a given weed.

The extrapolation from competition experiments and field surveys to economic thresholds is in any case difficult. Marshall's work (1988, see above) warns that many samples are required before the mean density can be estimated with any precision within an actual field. Also, clumping affects the crop yield/weed density relationship. The more clumped the weeds, the greater the intraspecific competition between them. This has the effect of reducing the overall yield loss of the crop, since the damage is concentrated in some areas but the less weedy areas more than make up for it (e.g., Brain and Cousens, 1990). Maps of fields, perhaps using balloons or model planes, may be needed for accurate estimates of control threshold (Thornton et al., 1990). Until we understand variation at this scale, our threshold values can act as imprecise guides at best.

The Temporal Dimension

Just as weed infestations can be studied at different spatial scales, so they can be studied at different time scales. Also, short-term effects can be integrated to have longer-term consequences; to use the jargon, there are emergent properties from one time scale to another.

The human-imposed variation occurs most obviously at the cropping cycle; the management between crops having a very large influence on weed populations. Weed research tends to reflect this, looking at the development of the weed/crop stand over a full season. The practical problem is that research results tend to be acquired after harvest, whereas often the farmer needs to be able to extrapolate to harvest from an earlier point in time. One important emergent property is therefore that the results of weed/crop competition can often be accounted for largely by differences in emergence time and growth at the start of the season. Simulation modeling by Spitters and Aerts (1983) strongly suggests that even a 12-h difference in emergence times between weed and crop can affect eventual yields. There are many field studies showing that the time of emergence of weeds and time of removal can be used to suggest optimal times for weed control (e.g., Nieto et al., 1968; Weaver, 1984). The effects of emergence time and density complement each other (Firbank et al., 1985), and can be assessed together by estimating weed biomass early in the season. This can give a good prediction of eventual yield loss (Kropff, 1988), although it can, of course, be overridden by subsequent events — a drought, a harsh winter, or an application of herbicide. Temporal variation can become confounded with spatial variation: for example, Firbank et al. (1990) suggested that a major source of spatial variation in weed/crop competition between sites was the result of different levels of waterlogging on the sites, affecting the emergence times of the weeds relative to the crop.

The performance of weeds in one season can have a large effect on numbers in furture seasons, a factor which needs to be taken into account when considering threshold weed densities. The reductionist view requires that estimates are made for the probabilities of survival of each stage of the weed's life cycle; these can be coupled with estimates of seed production and the behavior of the seed bank to produce a population model; spatial patterns and the dispersion effects are ignored (e.g., Watkinson, 1981; Mortimer, 1983; Cousens et al., 1987). Such models cannot give very precise estimates into the future because of the sensitivity of these probabilities to weather, details of the management, and so on (Firbank et al., 1985). Such random variation, coupled with chaotic effects, can lead in principle to weed population dynamics which are very different from those suggested by classical weed population models (Firbank, 1989).

There are several ways of dealing with this problem. The easiest way is to accept that such models act as guides, rather than as precise forecasts. Putwain and Mortimer (1989) have modeled the spread of herbicide-resistant weeds using this approach. If this is all that is required, then the level of detail offered by standard population models is too great. Instead of trying to model numbers of weed plants, one could consider ranking abundances of different weeds, or to model whether weed species are likely to increase or decrease under given situations. As less detailed measures of weed abundance tend to more stable (Rahel, 1990), the resulting models should be more robust.

A second approach is adopt stochastic models, models which incorporate a degree of random variation from year to year. The intention is not so much to produce precise forecasts, but rather to establish the risk of extinction or outbreak of species. Models have been developed by conservation biologists (e.g., Soulé, 1987) which could be easily adapted to weed populations, but the problem is that the data on year-to-year variation in the weed's population parameters are needed. There are very few long-term experiments to provide these data (Cousens et al., 1987), and we know of none which give the detail and range of weed densities needed to allow this approach to be used with confidence.

The third approach is more pragmatic, and is based on time-series models rather than on formal population models. The idea is to study trends from past data to suggest what may happen in the future. For example, Firbank's (1991) analysis of some of the data from the classic Broadbalk experiment at Rothamsted indicated that different weed species show different degrees of variation in numbers from year to year under the same management strategy. There is here the merest hint of useful generalization that some weed species are more easily forecasted than others.

As the timescale increases, the value of population models becomes weaker, and the effects of larger-scale factors become more apparent. These may involve changing patterns of farming, and even changing climate. These processes may seem to be too slow to be readily observed, but there are long-term data from other sources — local flora accounts, records and remains of seed impurities of seed corn, herbarium records, etc., which can be used to infer long-term changes in at least some weed species (e.g., Firbank, 1988).

The Dimension of Ecological Complexity

In most of this chapter so far, we have been discussing weed species in isolation from all other factors except the crop, the abiotic environment, and the farmer. We have concentrated on studies which deal with only one weed species at a time. In adopting this stance, we are merely following widely accepted practice, practice which has been encouraged by the specialization of disciplines within agroecology. Yet here too are issues of scale. Studies designed to investigate the behavior of a single species in isolation cannot be expected to yield an understanding of species interactions.

Perhaps one should consider the behavior of the whole weed community, rather than concentrated on single species. At a broad scale, this would be valuable; after all, changing farming practices aimed at attacking certain weeds may end up by promoting others — the increase of grass weeds in cereals in Europe attests to this. At the scale of the individual field, the benefits may be less substantial. If the different weed species are in separate clumps, they will interact little with other (Czaran and Bartha, 1989), and so it seems reasonable to consider them separately. We need more experiments to decide this issue.

The weed plants, together with the crop, form the autotrophic base of the cereal ecosystem. While they do interact with each other directly (via competition), they also have indirect interactions via herbivores or pathogens which should also be considered. These effects may be negative or positive. Altieri (1988) stresses the positive effects of weeds on crops, and lists many examples of cases where weeds have resulted in improved biological control of insect pests. This may result from encouraging the buildup of predators or parasitoids, or from the weed being a preferred host plant to a pest. In some cases the 'weed' may be planted deliberately as a trap crop (Hokkanen, 1991).

Even if the effects of the weeds upon crop yield are still negative, in the wider context of managing the whole agroecosystem, it may still be sound practice to allow weed populations to persist. Experiences from the 12-year comparison between conventional and integrated farming at Lautenbach have indicated that while weed levels are typically higher on the integrated plots, the integrated system is as a whole more sustainable and usually more profitable (El Titi, 1991). There are frequent suggestions that more diverse farming systems have fewer pest and weed problems in general than simpler systems, but more information is required (Francis, 1989).

THE OBJECTIVES OF WEED ECOLOGY

The study of weeds in small-scale experiments has been extremely important in developing our understanding of the building blocks of weed ecology. However, just as architecture does not confine itself to the study of bricks, so weed ecology must encompass larger scales and interactions. In particular, if the objective is to help the farmer, then larger-scale studies are essential.

The importance of the farm-scale experiment is clear by comparing Figure 1, which indicates typical spatial and temporal scales of experimental studies, with Figure 2, which shows the information desired by farmers, their advisors, and those involved in managing farming. The author believes that these people have few interests in small-scale effects, leaving the top and left of the diagram (so crowded in Figure 1) conspicuously empty. Even the yield loss assessments, so beloved of students of weed competition such as the author, are required for areas around an order of magnitude greater than that of typical competition field experiments.

Increasing attention is being paid to farm-scale experiments in the U.K. and elsewhere as ways of investigating farm-scale effects (e.g., Greig-Smith, 1991). However, these are expensive and not without problems (see below). Studies at larger scales, looking at the development of resistance to herbicides for example, are difficult to carry out experimentally at the scale of the response; simulation studies may prove particularly valuable here. However, we should beware of models which are so oversimplified that they fail to capture the essence of the system. Population models which ignore variation in space and time may be helpful tools for some purposes but certainly not for detailed prediction.

A World of Variation

To human eyes, the farming landscape in the U.K. is a patchwork of fields, hedges, lanes, roads, woods, and so on. We tend to interpret this as an essentially flat surface, which may acquire a third dimension as it rolls across hills, or as the odd tree pokes through. We often recognize greatest variation between fields

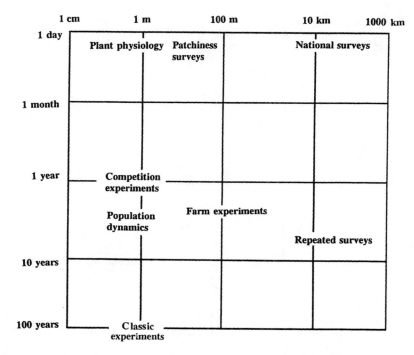

Figure 1. Schematic diagram of the scales of space and time of typical investigations into weeds. The scales are clearly indicative only. (Redrawn from Firbank, L. G. [1991]. *The Ecology of Temperate Cereal Fields,* L. G. Firbank, N. Carter, J. F. Darbyshire and G. R. Potts [Eds.], Blackwell Scientific, Oxford, 209–231. With permission.)

(if they are sown with different crops), and tend to regard variation within fields as small in contrast. However, reliance on a human-scale viewpoint imposes a strong bias on our understanding of the agricultural ecosystem.

For example, we might view the obvious difference between two fields as the difference between the crops. For different weed species, it may be distance from the hedgerow and its characteristics that are more important, or whether the crop was sown in the spring or in the winter, or the type of cultivation or soil type. Some of these effects are field scale, some more local. These factors help explain the importance of field size and cropping diversity on pest species (Altieri, 1988).

The very landscape differs between species. A field that is to us flat and featureless can be mountainous to smaller species, and show rich variety. At last, we can describe these changing perceptions mathematically in terms of fractal dimensions (e.g., Turner, 1989), but we have yet to incorporate these ideas into our understanding of agricultural systems (Macdonald and Smith, 1991). However, it should be clear that the same experiment may detect averaged-out, synoptic-level behavior of some species (Collembola and microbes, for example) and would be too small to detect any effects from others (birds of

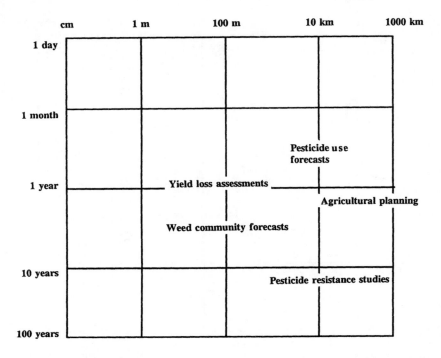

Figure 2. Schematic diagram of the scales of space and time of information required by farmers and planners about weeds. The scales are clearly indicative only. (Redrawn from Firbank, L. G. [1991]. *The Ecology of Temperate Cereal Fields*, L. G. Firbank, N. Carter, J. F. Darbyshire and G. R. Potts [Eds.], Blackwell Scientific, Oxford, 209–231. With permission.)

prey). The ability to generalize the results of any experimental depends upon which species we are looking at.

Nor should we neglect different perceptions of time. The changing patterns of species of microbes on the surface of the leaves of a crop display a remarkable succession within a season which is quite comparable to much longer-lasting successions of old fields (Blakeman, 1991). Again, a given sampling program will give a synoptic view of some species yet will tell us hardly anything of value for others.

Ideally, one should identify the scales of maximum variability in population levels for a given species of interest. These findings can then be interpreted in the light of other knowledge. For example, if a species varies greatly at the level of the area of a field, then one can assume that differences between fields are important, and that the management of the fields may well have a strong effect. The same conclusion would come from a species which shows strong year-to-year fluctuations which correlate better with management than with the weather. Time series and pattern analysis can help us find these scales of maximum variation, but such methods require intensive sampling over a large area or a long period of time. It is no surprise that these methods have rarely been applied to the agroecosystem.

CONCLUSION

To summarize, much information on weeds, and on the agroecosystem in general, is collected at scales of time, space, and complexity of convenience to experimenters. These scales are often smaller than those of interest to farmers, and it is often not easy to extrapolate from the experimental scale to the farm scale. The increasing use of farm experimental systems recognizes this problem, but the high degree of variation between fields and areas, coupled with the host of management options to be explored, make the kind of replication acceptable in small plot experiments virtually impossible on whole farms. Finally, what may be an appropriate scale of experiment or monitoring exercise for one species may be woefully inadequate for others — potentially a major problem when species interactions are studied, especially between levels of food webs. The idea that one size of experiment should be pursued at the expense of all others cannot be defended. Also, despite the increasing sophistication of methods of looking at scale and pattern (e.g., Allen and Starr, 1982; Turner, 1989; Weins, 1989), we must conclude that a complete understanding of agroecosystems is beyond us.

There are two ways out of this problem. The first is that clear research priorities are required. Research may be into whole systems, or particular species. The studies should be designed with the ecology of the species in mind, and the level of detail required. The difficulty will always remain that such research may come too late to be of use, so some level of general monitoring is required to detect changes in the agroecosystem as they happen (e.g., Woiwod, 1991).

The second solution is less obvious. The problem for agroecologists is the shortage of resources to investigate the full range of conditions which are encountered in the field. We have little faith that our farm experiments provide sufficient samples for our needs. Yet every time a farmer plants a crop in a field, an experiment takes place (Woiwod, 1991). If the objective of research in agroecology is to improve the management of our landscape, then the farmers must learn from each other as well as from 'experts' who reply on a limited database. Farmers clubs dedicated to sharing information may turn out to be the best safeguard the agroecosystem could possibly have.

REFERENCES

Aebischer, N. (1991). Twenty years of monitoring invertebrates and weeds in cereal fields in Sussex. *The Ecology of Temperate Cereal Fields*, L. G. Firbank, N. Carter, J. F. Darbyshire and G. R. Potts (Eds.), Blackwell Scientific, Oxford, 305–331.

Allen, T. F. H. and Starr, T. B. (1982). *Hierachy Theory: Perspectives for Ecological Complexity*. University of Chicago Press, Chicago, 310.

Altieri, M. A. (1988). *Agroecology: The Scientific Basis of Alternative Agriculture*, Westview Press, Boulder, CO, 227.

Begon, M., Harper, J. L. and Townsend, C. R. (1990). *Ecology*, 2nd ed., Blackwell Scientific, Oxford, 945.

Blakeman, (1991). Microbial interactions in the phylloplane. *The Ecology of Temperate Cereal Fields*, L. G. Firbank, N. Carter, J. F. Darbyshire, and G. R. Potts (Eds.), Blackwell Scientific, Oxford, 171–191.

Brain, P. and Cousens, R. (1990). The effect of weed distribution on predictions of yield loss. *J. Appl. Ecol.*, 27:735–742.

Carney, H. J. (1990). On competition and the integration of population, community and ecosystem studies. II. Replies to Fenchel, Allen and Hoekstra. *Functional Ecol.*, 4:127–128.

Cousens, R. (1985). A simple model relating yield loss to weed density. *Ann. Appl. Biol.*, 107:239–252.

Cousens, R., Moss, S. R., Cussans, G. W. and Wilson, B. J. (1987). Modelling weed populations in cereals. *Rev. Weed Sci.*, 3:93–112.

Cousens, R., Firbank, L. G., Mortimer, A. M. and Smith, R. R. (1988). Variability in the relationship between crop yield and weed density for winter wheat and *Bromus sterilis*. *J. Appl. Ecol.*, 25:1033–1044.

Czaran, T. and Bartha, S. (1989). The effect of spatial pattern on community dynamics; a comparison of simulated and field data. *Vegetatio*, 83:229–239.

Eberhardt, L. L. and Thomas, J. M. (1991). Designing environmental field studies. *Ecol. Monogr.*, 61:33–73.

El Titi, A. (1991). The Lautenbach project 1978–89: integrated wheat production on a commercial arable farm in Southwest Germany. *The Ecology of Temperate Cereal Fields*, L. G. Firbank, N. Carter, J. F. Darbyshire and G. R. Potts (Eds.), Blackwell Scientific, Oxford, 399–411.

Firbank, L. G. (1988). Biological flora of the British Isles No. 165, *Agrostemma githago* L. *J. Ecol.*, 76:1232–1246.

Firbank, L. G. (1989). Forecasting weed infestations — the desirable and the possible. Brighton Crop Proection Conf. — Weeds 1989. British Crop Protection Council, Farmham, 567–572.

Firbank, L. G. (1991). Interactions between weeds and crops. *The Ecology of Temperate Cereal Fields*, L. G. Firbank, N. Carter, J. F. Darbyshire and G. R. Potts (Eds.), Blackwell Scientific, Oxford, 209–231.

Firbank, L. G. and Watkinson, A. R. (1985a). A model of interference within plant monocultures. *J. Theor. Biol.*, 116:291–311.

Firbank, L. G. and Watkinson, A. R. (1985b). On the analysis of competition within two-species mixtures of plants. *J. Appl. Ecol.*, 22:503–517.

Firbank, L. G., Manlove, R. J., Mortimer, A. M. Putwain, P. D. (1984). The management of grass weeds in cereal crops, a population biology approach. Proc. 7th Int. Symp. Weed Biol. Ecol. Syst., 375–384.

Firbank, L. G. Mortimer, A. M. and Putwain, P. D. (1985). *Bromus sterilis* in winter wheat: a test of a predictive population model. *Aspects Appl. Biol.*, 9:59–66.

Firbank, L. G., Cousens, R., Mortimer, A. M. and Smith, R. R. (1990). Effects of soil type on crop yield — weed density relationships. *J. Appl. Ecol.*, 27:308–318.

Francis, C. A. (1989). Biological efficiencies of multiple-cropping systems. *Adv. Agron.*, 42:1–42.

Gill, G. S., Poole, M. L. and Holmes, J. E. (1987). Competition between wheat and brome grass in Western Australia. *Aust. J. Exp. Agric.*, 27:291–294.
Greig-Smith, P. W. (1991). The Boxworth experience: effects of pesticides on the flora and fauna of cereal fields. *The Ecology of Temperate Cereal Fields*, L. G. Firbank, N. Carter, J. F. Darbyshire and G. R. Potts (Eds.), Blackwell Scientific, Oxford, 333–371.
Harper, J. L. (1977). *The Population Biology of Plants*. Academic Press, New York, 892.
Hokkanen, H. M. T. (1991). Trap-cropping in pest management. *Ann. Rev. Entomol.*, 36:119–138.
Kershaw, K. A. and Looney, J. H. H. (1985). *Quantitative and Dynamic Plant Ecology*, 3rd ed., Edward Arnold, London, 282.
Koestler, A. (1967). *The Ghost in the Machine*, Hutchinson, London, 383.
Kropff, M. J. (1988). Modelling the effects of weeds on crop production. *Weed Res.*, 28:465–471.
Lovelock, J. E. (1979). *Gaia: a New Look at Life on Earth*, Oxford University Press, Oxford, 157.
Macdonald, D. W. and Smith, H. (1991). New perspectives on agroecology: between theory and practice in the agricultural ecosystem. *The Ecology of Temperate Cereal Fields*, L. G. Firbank, N. Carter, J. F. Darbyshire and G. R. Potts (Eds.), Blackwell Scientific, Oxford, 413–448.
Marshall, E. J. P. (1988). Field-scale estimates of grass weed populations in arable land. *Weed Res.*, 28:191–198.
Marshall, E. J. P. (1989). Distribution patterns of plants associated with arable field edges. *J. Appl. Ecol.*, 26:247–257.
Mortimer, A. M. (1983). On weed demography. Recent Advances in Weed Research, W. W. Fletcher (Ed.), Commonwealth Agricultural Bureau, Farnham Royal, 3–40.
Putwain, P. D. and Mortimer, A. M. (1989). The resistance of weeds to herbicides: rational approaches for containment of a growing problem. Brighton Crop Protection Conf. — Weeds 1989. British Crop Protection Council, Farnham, 285–294.
Nieto, J., Brondo, M. A. and Gonzalez, J. T. (1968). Critical periods of the crop growth cycle for competition from weeds. *Pans(C)*, 14:159–166.
O'Niell, R. V. (1989). Perspectives in hierarchy and scale. *Perspectives in Ecological Theory*, J. Roughgarden, R. M. May and S. A. Levin (Eds.), Princeton University Press, Princeton, NJ, 140–156.
Rahel, F. J. (1990). The hierarchical nature of community persistence: a problem of scale. *Am. Nat.*, 136:328–344.
Sheail, J. (1987). *Seventy Five Years in Ecology: The British Ecological Society*, Blackwell Scientific, Oxford, 301.
Soulé, M. E. (Ed.) (1987). *Viable Populations for Conservation*, Cambridge University Press, Cambridge, 189.
Spitters, C. J. T. and Aerts, R. (1983). Simulation of competition for light and water in crop-weed associations. *Aspects Appl. Biol.*, 4:467–483.
Streibig, J. C., Conbellack, J. H., Pritchard, G. H. and Richardson, R. G. (1989). Estimation of thresholds for weed control in Australian cereals. *Weed Res.*, 29:117–126.
Thornton, P. K., Fawcett, R. H., Dent, J. B. and Perkins, J. J. (1990). Spatial weed distribution and economic thresholds for weed control. *Crop Prot.*, 9:337–342.

Turner, M. G. (1989). Landscape ecology: the effect of pattern on process. *Ann. Rev. Ecol. Syst.*, 20:171–197.

Watkinson, A. R. (1981). Interference in pure and mixed populations of *Agrostemma githago*. *J. Appl. Ecol.*, 18:967–976.

Weaver, S. E. (1984). Critical period of weed competition in three vegetable crops in relation to management practices. *Weed Res.*, 24:317–325.

Weins, J. A. (1989). Spatial scaling in ecology. *Functional Ecol.*, 3:385–397.

Woiwod, I. P. (1991). The ecological importance of long term synoptic modelling. *The Ecology of Temperate Cereal Fields*, L. G. Firbank, N. Carter, J. F. Darbyshire and G. R. Potts (Eds.), Blackwell Scientific, Oxford, 275–304.

10 Total Species Number as a Criterion for Conservation of Hay Meadows

M. H. LOSVIK

Abstract: *The total species number of 130 hay meadow sites in western Norway was split into 3 different groups of species, namely: trivial species (with a wide distribution in grasslands), forest species, and traditional species (considered generally to indicate traditional management). Regression analysis showed that the number of trivial species is independent of quantities of fertilizer in the mesotrophic meadow data set, while total species number and number of traditional species is negatively correlated to quantities of fertilizer used. Regression analysis of a subset indicated that the number of trivial species may be positively correlated to the quantities of fertilizer used. In such cases, the use of total species numbers as a criterion for hay meadows may obscure the fact that the traditional species are often dependent on small quantities of NKP in the soil.*

INTRODUCTION

The aim of this study was to search for a group of species which might be used as indicators of traditionally managed hay meadows in western Norway.

Usually, the number of species is higher in unfertilized mesotrophic meadows than in well-fertilized meadows (Traczyk and Kotowska, 1976; Traczyk et al., 1976; Maarel, 1971; Zoller and Bischof, 1980). Also, it is well known which species are present in unfertilized meadows and in well fertilized ones, and how species composition may change with increased fertilization (Foerster, 1964; Thurston, 1969; Ellenberg, 1978; Elbersee et al., 1983; Lundekvam and Gauslaa, 1986; Losvik, 1988).

However, it has been observed that some well fertilized meadows may have significant numbers of trivial species, while unfertilized meadows, on the other hand, may have low species diversity, even if the species represented are quite rare. Such meadows may be interesting for conservation purposes and it is therefore useful to understand the processes involved.

Traditionally, in western Norwegian hay meadows the grass is cut rather late in summer, from the middle of July onward, but the meadows are neither fertilized nor manured, except for manuring by grazing animals in spring and autumn

(Visted and Stigum, 1951; Norges landbruksøkonomiske institutt, 1955; Sølvberg, 1976; Byrkjeland, 1958; Nesheim, 1983; Losvik, 1988).

It is therefore necessary to establish which species are recorded more or less exclusively in unfertilized hay meadows, as opposed to those which are more common both in unfertilized and in well fertilized hay meadows. The former group of species would be negatively correlated to the quantities of fertilizer used, while the latter are likely to be rather independent of fertilizing.

METHODS

To test this hypothesis, species were recorded from 130 plots of 16 m^2 each, taken from mesotrophic hay meadows within Molinio-Arrhenatheretea, order Arrhenatheretalia, in western Norway (Losvik, 1988). All species were divided into three different species groups, namely; trivial species, forest species, and traditional species.

The group of trivial species comprised some NKP-demanding species like *Ranunculus repens* and *Poa trivialis,* and others with a wide ecological spectrum within mesotrophic meadows, e.g., *Rumex acetosa, Ranunculus acris, Agrostis capillaris,* and *Anthoxanthum odoratum.* Altogether, 39 species were chosen (see Appendix). These are considered to be nonindicators or only weak indicators of traditional management, even if several of them are important constituents of traditionally managed hay meadows.

The forest species group comprised species with their main distribution in forests, like *Fragaria vesca, Luzula pilosa,* and *Anemone nemorosa* for a total of 25 species in addition to mosses (except *Rhytidiadelphus squarrosus*), scrubs, and trees (see Appendix).

Species within these two groups were subtracted from the total number of species in each plot. The rest of the species were regarded as indicators of traditionally managed meadows in a broad sense and here considered as traditional species (see Appendix). The group comprised species with rather low NKP demands; several are tolerants of stress and/or grazing. Examples of traditional species are *Conopodium majus, Leontodon autumnalis, Plantago lanceolata, Holcus lanatus, Leucanthemum vulgare, Rhinanthus minor,* and *Pimpinella saxifraga.*

Regression analysis was carried out using quantities of fertilizer and number of trivial species and traditional species in each plot. Three different levels of fertilizer use were examined: no fertilizing (1), light fertilizing (2), and medium fertilizing (3) (defined as 70–120 kg N, 15–50 kg P, and 75–100 kg K/ha/yr). Data concerning the levels of fertilizer for each plot were obtained from the farmers. As the history of the hay meadows is not considered, noise may be present in the data material, caused by the slow reaction of some species to possible recent changes in quantities of fertilizer applied.

RESULTS

In the regression analysis the three classes of fertilization were plotted on the x-axis and the number of species on the y-axis (Figure 1). The data showed that the trivial species, as expected, were not correlated to doses of fertilizer ($r = -0.03$). Forest species showed a weak negative correlation ($r = -0.18$, $p < 0.05$).

The traditional meadow species, however, were negatively correlated to the quantity of fertilizer used ($r = -0.3$, $p < 0.01$), and seem to account for most of the negative correlation of the total number of species ($r = -0.3$, $p < 0.01$).

The data set comprised eight different plant communities. In one of these subsets, containing 25 plots, the number of trivial species even showed a weak positive correlation to doses of fertilizer used ($r = 0.14$, not significant), while neither the number of traditional meadow species nor the forest species were correlated to the quantities of fertilizer used ($r = -0.08$ and 0.04, respectively). The total species number was also independent of differences in quantities of fertilizer applied ($r = 0.05$). This type of community was considered interesting for conservation purposes, with high numbers of species compared to the other communities.

CONCLUSION

Even if the total number of species is lower in well-fertilized hay meadows than in unfertilized ones, in most of these data sets a large proportion of the species present appear to occur quite independently of fertilizing or may even be positively correlated to quantities of fertilizer used in up to medium doses. Another group of species, considered generally to indicate traditional management, seem to account for most of the negative correlation of the total species number with the fertilizer level.

The concept of total species number in grasslands (e.g., Silvertown, 1980; During and Willems, 1984; Collins and Barber, 1985) should be reconsidered. In comparing total species numbers in plots from different meadows, the presence of significant indicators or rare species may be obscured by indifferent species. The analysis of the subset of plots showed the importance of restricting the use of fertilizer to retain the rarer species in the hay meadows, as the number of trivial species may even increase with higher fertilizer levels. Thus, total species numbers should only be used to complement the number of characteristic species as, for example, Willems (1983) has done in comparing grassland diversity.

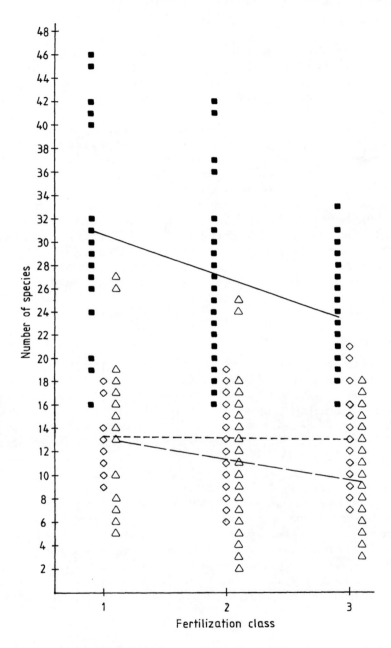

Figure 1. Relationship between quantity of fertilizer (classes 1–3, see text) and number of trivial species ($\diamond$, ------, r = -0.03); traditional species ($\triangle$, — —, r = -0.3, $p < 0.01$); and total number of species ($\blacksquare$, ———, r = -0.3, $p < 0.01$).

APPENDIX Groups of Species

1. Trivial species

Rhytidiadelphus squarrosus
Rumex acetosa
Poa pratensis
Taraxacum vulgaris coll.
Trifolium repens
Festuca pratensis
Alopecurus geniculatus
Caltha palustris
Lychnis flos cuculi
Viola palustris
Ranunculus repens
Anthriscus sylvestris
Veronica serpyllifolia
Deschampsia flexuosa
Stellaria graminea
Deschampsia cespitosa
Rumex acetosella
Silene vulgaris
Lolium perenne
Anthoxanthum odoratum

Ranunculus acris
Festuca rubra
Cerastium fontanum
Trifolium pratense
Dactylis glomerata
Alopecurus pratensis
Montia fontana
Cardamine pratensis
Myosotis cespitosa
Alchemilla vulgaris coll.
Phleum pratense
Poa trivialis
Bellis perenne
Festuca ovina
Nardus stricta
Cirsium palustre
Sagina procumbens
Rumex longifolia
Agrostis capillaris

2. Forest species[a]

Anemone nemorosa
Carex panicea
Vaccinium myrtillus
Holcus mollis
Luzula sudetica
Athyrium felix-femina
Filipendula ulmaria
Valeriana sambucifolia
Oxalis acetosella
Ranunculus ficaria
Trientalis europaea
Juncus filiformis
Melandrium rubrum

Carex leporina
Carex pulicaris
Carex nigra
Luzula pilosa
Angelica sylvestris
Epilobium angustifolium
Galium mollugo
Fragaria vesca
Ranunculus auricomus
Solidago virgaurea
Viola riviniana
Juncus effusus

3. Traditional species

Conopodium majus
Vicia sepium
Leontodon autumnalis
Achillea ptarmica
Bromus hordeaceus
Hypericum maculatum
Carex pallescens

Avenula pubescens
Polygonum viviparum
Ajuga pyramidalis
Festuca vivipara
Lathyrus montanus
Carex pilulifera
Galium saxatilis

3. Traditional species (continued)

Luzula campestris	Vicia cracca
Plantago lanceolata	Galium boreale
Holcus lanatus	Achillea millefolium
Galium uliginosum	Knautia arvensis
Carum carvi	Linum catharticum
Briza media	Euphrasia stricta
Centaurea nigra	Leucanthemum vulgare
Danthonia decumbens	Pimpinella saxifraga
Centaurea jacea	Prunella vulgaris
Rhinanthus minor	Lotus corniculatus
Hieracium vulgatum	Luzula multiflora
Succisa pratensis	Veronica chamaedrys
Geranium sylvaticum	Hieracium umbellatum
Veronica officinalis	Potentilla erecta
Campanula rotundifolia	Hypochoeris radicata
Trifolium dubium	Myosotis ramosissima
Polygala vulgaris	Cynosurus cristatus
Plathanthera chlorantha	

a In addition, recorded mosses, except *Rhytidiadelphus squarrosus*, and the recorded trees and scrubs are considered to be forest species.

REFERENCES

Byrkjeland, J. (1958). Husdyrbruket i Hordaland gjennom 100 år. Hordaland landbruksselskap, Bergen, 64–93.

Collins, S. L. and Barber, S. C. (1985). Effects of disturbance and diversity in mixed-grass prairie. *Vegetatio*, 64:87–94.

During, H. J. and Willems, J. H. (1984). Diversity models applied to a chalk grassland. *Vegetatio*, 57:103–114.

Elbersee, W. T., van den Bergh, J. P. and Dirven, J. G. P. (1983). Effects of use and mineral supply on the botanical composition and yield of old grassland on heavy-clay soil. *Neth. J. Agric. Sci.*, 31:63–88.

Ellenberg, H. (1978). *Vegetation Mitteleuropas mit den Alpen in ökologischer Sicht.* Ulmer, Stuttgart, 981.

Foerster, E. (1964). Zur Systematischen Stellung artenarmer *Lolium*-Weiden. *Pflanzensoziologische Systematik*, R. Tüxen (Ed.), Bericht. Int. Symp. Stolzenau/Weser Int. Vereig. Vegetat., Dr. W. Junk N.V., Den Haag, 183–190.

Losvik, M. H. (1988). Phytosociology and ecology of old hay meadows in Hordaland, western Norway in relation to management. *Vegetatio*, 78:157–187.

Lundekvam, H. E. and Gauslaa, Y. (1986). Phytosociology and ecology of mown grasslands in western Norway. *Meld. Norg. LandbrHøgsk.*, 65(22):1–26.

Nesheim, L. (1983). Avlingsnivå og kvalitet på eldre eng i Nordland. *Rapp. Norges Landbr. Vitensk. Forskningsråd*, 481:1–14.

Norges Landbruksøkonomiske Institutt (1955). Kystjordbruket. *Særmeldinger* 8:1–166.

Silvertown, J. (1980). The dynamics of a grassland ecosystem: botanical equilibrium in the Park grass experiment. *J. Appl. Ecol.*, 17:491–504.

Sølvberg, I. Ø. (1976). Driftsmåter i vestnorsk jordbruk ca 600–1350. Universitetsforlaget, Oslo, 197 pp.
Thurston, J. M. (1969). The effect of liming and fertilizers on the botanical composition of permanent grassland, and on the yield of hay. *Ecological Aspects of the Mineral Nutrition of Plants*, I. H. Rorison (Ed.), Symp. Br. Ecol. Soc., Blackwell Scientific, Oxford, 3–11.
Traczyk, T. and Kotowska, J. (1976). The effect of mineral fertilization on plant succession of a meadow. *Pol. Ecol. Stud.*, 2:75–84.
Traczyk, T., Tacczyk, H. and Pasternak, D. (1976). The influence of intensive mineral fertilization on the yield and floral composition of meadows. *Pol. Ecol. Stud.*, 2:39–47.
Willems, J. H. (1983). Species composition and above ground phytomass in a chalk grassland with different management. *Vegetatio*, 52:171–180.
Zoller, H. and Bischof, N. (1980). Stufen der Kulturintensität und ihr Einfluss auf Artenzahl und Artengefüge der Vegetation. *Phytocoenologia*, 7:35–51.
Maarel, van der, E. (1971). Plant species diversity in relation to management. The Scientific Management of Animal and Plant Communities For Conservation, *Symp. Br. Ecol. Soc.*, E. Duffey and A. S. Watt (Eds.), 11:45–63.
Visted, K. and Stigum, H. (1951). *Vår Gamle Bondekultur.* I. J. W. Cappelens Forlag, Oslo, 240.

11 The Diversity of Agricultural Practices and Landscape Dynamics: The Case of a Hill Region in the Southwest of France

NICOLE SAUGET
GÉRARD BALENT

Abstract: *Farming activities can manage, neglect, or breakdown agroecosystems. In this project, we studied the diversity of the impact of agricultural practices on landscape dynamics. The spatial coexistence and contiguity of different practices, and the uncertainty about the local scale of social and ecological heterogeneity forced us to study an entire territory. The area is located in southwest France. There we investigated the relationship between the diversity of the agricultural practices (at field and farm levels) and the position of landscape units along ecological gradients. For comparative purposes, we used a method that allowed us to link these three different patterns. The results show the impact of the practices which characterize each field, and the importance, at a higher level, of practices involving whole-systems which correspond to different family/farm systems.*

LANDSCAPE ECOLOGY AND THE DIVERSITY OF THE FARMING STYLES

Golley (1988) stressed the importance of involving the human element in landscape analysis. The impact of human action on the ecosystem is more often taken into consideration as a disturbance rather than as a management factor. But when farming is present, it should be an urgent concern to account for what takes place at the agroecosystem level, and the diverse effects of agricultural practices on the landscape: people can manage, neglect, or cause decline in patterns through their farming and husbandry activities.

Gulinck (1986) suggested that there are two main principles of study in landscape ecology, although they are not equally used: the first considers the way in which agroecosystems are simply situated in the landscape and how they spatially interact with the other components of the environment, and the second focuses on how man actively manages agroecosystems and structures the landscape according to different uses of natural and human resources.

The second way enables us to study the diversity in the farmers' influence on the landscape. Authors from various disciplines (among them Deleage et al., 1978; Buttel, 1979, 1984; Godelier, 1979, 1984; Pimentel, 1979, 1990; Lemonnier, 1983; Norgaard, 1986; Golley, 1988) have emphasized that the interaction between agricultural activity and ecosystems can be qualitatively different according to areas, times, and cultures, as may also be the more global human/landscape relationship. We consider that in one society, in one district, different modes of this interaction can coexist at the same moment and need to be studied in order to evaluate their sustainability.

In France, for instance, social evolution has generated throughout history a marked diversity among farmers' families and farming methods; differences which were reinforced due to local conditions and the various types of enterprises (Jollivet, 1974, 1984, 1988; Deleage et al., 1979; Sauget, 1979; Remy, 1982; Eizner, 1985; Mathieu and Dubosq, 1985; Braudel, 1986; Sauget and Maresca, 1986). During the last few decades, with the retirement of many traditional farmers and the disappearance of a number of small farms as a result of socioeconomic constraints, the general trend towards high productivity of agricultural work as well as the concentration and specialization of farms has tended to reduce this diversity. Other factors, such as the migration of farmers from one region to another, the arrival of newcomers in agriculture, and the interest in organic farming have also contributed locally to its renewal. Diversity in the farming approaches and methods, and heterogeneity in the landscape are changing in scale, with different processes taking place in different regions.

The spatial coexistence of various types of farmers, and the uncertainty about the local scale of social and ecological heterogeneity, show the need for an interdisciplinary approach, and the necessity for local studies held on an entire territory. The rhythms in time of the various changes also need to be taken into consideration.

Specific Features of the Study

Hypotheses

Osty (1978) stated that "We consider as one system the package (whole) constituted by the farm, the farmer and his family", and the SAD (Agrarian Systems and Development) Department of INRA generally uses the concept of a "family/farm system". In our approach to the examination of the present processes, we assume that farming styles are whole systems, involving more than the systems of practices, technical systems, and farming systems separately. They express the constraints of several higher levels of cultural, social, and economic contexts, and include some aspects of the family way of life. Farming styles are the living expression of the family/farm systems: in present-day agriculture in France, various combinations of cultural backgrounds, social and family contexts, and farming systems remain, which induce certain practices and ecological impacts (Sauget and Maresca, 1986). Some are more sustainable than others in social, or in ecological and landscape quality terms.

Landscape ecology deals with landscape dynamics on a continuous spatial basis, which is necessary if one seeks to account for the effects of agricultural practices. In the same way, and because it is necessary to account for particular and real processes in order to link specific practices with specific impacts, we needed also to get information about agricultural practices on a continuous spatial basis.

Our first hypothesis was that in a continuously variable rural territory, such as that of a village in southwestern France, a link can be found between farmers' practices at the field level and the landscape pattern revealed through ecological surveys using indicators such as bird numbers and species richness. The second hypothesis was that in such a territory farmers differ not only in the practices regarding a corn field or a hay meadow, for instance, but also in their overall farm/family system. We also wanted to test whether there was a link between these landscape and farm patterns.

The diagnostic assessment of ecological systems may rely on measurement of the diversity and changes in the composition and organization of plant or animal communities which are considered as ecological indicators. Thus, Balent and Courtiade (1992) have used an approach based on the exhaustive modeling of passerine bird communities along three ecological gradients. In the same district, we proposed studying on an exhaustive basis both the field practices and the farming systems of the people who use the land of a village.

On a given continuously variable and heterogenous territory, many of the differences in practice which determine the variations in impact are due to decisions and habits at the farm level. Others pertain to other levels: the village, the area under irrigation, and national or European levels. We have mainly concentrated on the farm level, where we collected information concerning the families, the systems of practices, and every field as a land use parcel. We also considered some higher levels such as control and constraint levels. In order to analyze in the future the coevolution of the farms and the landscapes, we assumed that the changes taking place at the upper constraint levels should be taken into consideration when scheduling new surveys.

The Case of Saint-André

The area where both farming practices and landscape ecology were studied exhaustively is the village of Saint-André, situated north of the district of Aurignac, in the Haute-Garonne department (Figure 1). The village (1827 ha) comprises several parts. First, two sets of hills on both sides of the river Nere which runs from west to east, and its 625-m-wide valley. On the south part, the hillside descends northwards to the Nere, with a 20% slope. It is mainly covered with woods and natural pastures, sometimes evolving into overgrown fallow land invaded by *Juniperus communis*. On the north, a set of parallel thalwegs are diversely used for pasture, meadow, and cereals, and rise up to a common crest which runs parallel to the river (de Ravignan, 1981; Remondière, 1982). Only a few patches of fallow land can be seen, and hedges have frequently been

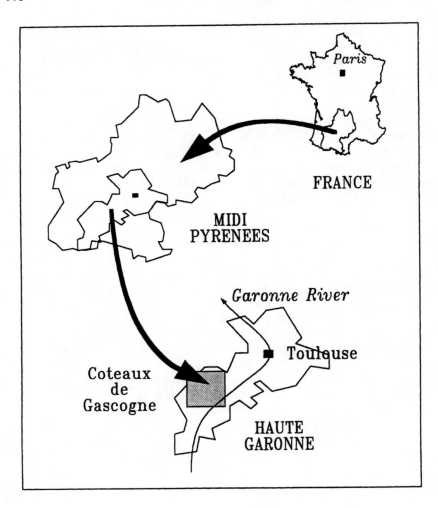

Figure 1. Location of the study area.

removed. There can be traces of soil damage where the slopes have been tilled, especially in the case of large fields with reduced vegetation cover.

The village is situated on one of the aforementioned crests, surrounded by very small fields; the rest of the houses are farms distributed on other crests, surrounded by their hillside fields, and possibly with some parcels in the valley. The valley is covered with pasture, forage crops, and sometimes cereals. The plots are generally very small and were delineated, at the time of our survey, by remnants of a hedgerow network perpendicular from the river, which was still lined with trees. North of the village, a third hill area carries mainly pastures lined by shrubs or woods, and fallow land is also present.

In Saint-André, as in the district as a whole, agriculture was traditionally devoted to mixed farming systems mainly oriented toward the production of veal (Perget, 1983; Prouzet, 1983). Subsequent development has irregularly and successively led to other enterprises. Milk production became more intensive during the 1970s, and increased until the EEC "quotas" were introduced in 1983. It is now more concentrated in rather large farms.

In the 1980s the trends have evolved in two opposite directions. Firstly, toward an intensive cultivation of corn and sunflower, both in the valleys and on certain slopes, with an acceleration in the removal of hedgerows. Secondly, by contrast, the number of cattle and sheep decreased locally, leading toward more extensive grazing use and development of overgrown fallow land. An important socioeconomic characteristic of the region is the importance of cooperative buying and use of machinery by farmers through the CUMA organization. It is generally considered to be efficient in slowing down the decrease in the number of local farmers.

Methodological Choices

We surveyed the farmers of Saint-André, collecting the data we needed in order to describe the family/farm systems, their relations with their contexts, and the resulting practices. The data concerning farming practices and the level of the various inputs were collected at the level of every farming plot. Then we selected the most significant ones for a first comparison with the appropriate ecological indicators (Table 1).

Regarding the ecological indicators, we had covered the territory of the commune with a grid with sides of 250 m, thus creating a network of landscape units for data gathering and analysis. A model provided the ordination of the landscape units along ecological gradients (Balent and Courtiade, 1992). It wastaken as a reference basis for positioning the landscape units. On the other hand, the practices and farming systems data were described at the field level and the farm level. These three series of data were connected by use of the network of landscape units. A projection of the center of every field surveyed, characterized by its spatial coordinates, was applied inside this network, each field remaining associated at the same time with the farm to which it belonged. This solution gave us a basis for a preliminary comparison regulated by the scale of the grid, and also for further investigation. We first established maps of the various inputs according to the grid structure and compared them with maps of the biological diversity.

Whereas this scale was fine enough to provide a broad picture of the landscape patterns, and the general use of each input in various parts of the village, it proved too coarse to evaluate the detail of the data we had gathered on the fields when we sought to relate them with each farm's whole system. This was still more so in the case of many small fields belonging to different farms. Therefore we changed scale by adding another network of landscape units, 16 times finer (2592 grid cells), and using this new grid in order to project the center of the

Table 1 Example of Inputs on the Different Fields (BA Farm)

Fields	Cadastral land parcels	Land Use	(Unit/ha/yr) N	P	K	(4 classes: 0 — 1 — 2 — 3) Herbicide	Insecticide	Fungicide	Manure	(hr/ha/yr) Tractor time
BA 1	222.224	Wood	—	—	—	—	—	—	—	—
BA 2	227.233	Barley	100	100	100	2	—	1	—	5
BA 3	223.221	Perm. meadow	50	30	30	—	—	—	—	5.30
BA 4	245.247	Wheat	150	100	100	2	—	—	—	6
BA 5	221	Silo	—	—	—	—	—	—	—	0.30
BA 6	221	Yard	—	—	—	—	—	—	—	—
BA 7	222	Sunflower	—	160	150	2	3	—	—	6
BA 8	285.295	Sunflower	—	175	160	2	3	—	—	6
BA 9	293.294	Temp. meadow	100	100	100	—	—	—	1	8
BA 10	225.226	House	—	—	—	—	—	—	—	—
Ba 11	81	Perm. meadow	—	—	—	—	—	—	—	4
BA 12	27 to 30	Perm. meadow	—	—	—	—	—	—	—	4
BA 13	30	Temp. meadow	150	100	100	—	2	—	—	8
BA 14	34	Temp. meadow	150	100	100	—	2	—	—	8
BA 15	210	Perm. meadow	50	—	—	—	—	—	—	3.30
BA 16	207.208	Perm. meadow	50	—	—	—	—	—	—	3.30

various fields and their associated practices by using the kriging algorithm of the grid routine in the Surfer Software (Golden Software, Golden, CO, 1990). For easier comparison, the avifauna pattern was divided on the new grid, although the data were gathered at the scale of the coarser one. We considered that we could use this type of comparison since the measurement, in the case of avifauna, deals with a more diffuse character than in the case of the farmer's practices.

Assuming that some explanation for perceived phenomena is to be found at the different scales where decisions occur, we applied the above results to the land farmed by the different families. We were thus able to look at the relation between the landscape and the farm patterns. In the meantime, the structure of the data on farms and farmer families enabled them to be considered as whole systems. We were then in a position to compare the data and see whether the complete systems were linked with ecological diversity.

Agricultural Practices and Landscape Dynamics on a Continuous Spatial Basis

A Comparison at Village Level

First we established maps of the various inputs according to the first grid structure, compared them to the maps of the landscape seen through the passerine bird species diversity, and showed that correlations did exist at the village level between these different patterns.

- The total number of birds pattern, as already shown (Balent, unpublished data), is not very different from the richness pattern. On the contrary, both are opposed to and almost complementary of the open landscape pattern (F1)* characterized through a particular composition of species.
- The mapping of nitrogen (N), herbicide, and tractor time inputs of the fields is largely opposite to that of the number of birds and richness of species.
- The open landscape character, on the other hand, appears to coincide with heavy inputs either of nitrogen, herbicide, or tractor time, separately or together.

A Finer Image Shows the Diversity of the Practices Among Farmers

In a second stage, use of the fine grid allowed us to exploit the precision obtained in the data relating to every farmer's practices. It enabled us to visualize more precisely the relationships between the different patterns and to focus on the way they corresponded to the different farms, hence to see how they might be linked not only with practices on a field but also with the overall individual farm enterprise:

- The link between the pattern of herbicide input and the pattern of the open landscape species composition is more pronounced in the central and western parts of the map.

---

* F1: ordination of landscape units along the first axis of CA, expressing different levels of the open landscape character (see Table 2).

- There is also a link between the herbicide, N, and to a certain extent P and K input patterns. All are connected with open landscape species composition pattern, the fertilizer and tractor time inputs, however, being less strongly connected with it. The organic manure spreading pattern, on the other hand, differs, and complements the chemical nitrogen pattern.

Removal of the hedgerows network and use of herbicide therefore determine the main landscape characteristics in the village, and are caused by the adoption of farming methods which have been considered "progressive" in France for many years. These results should be compared with the hedgerow network of the village.

When we look at the results on the new grid through the structure of the farms pattern, differences between the farm territories appear in diversity in the use of inputs and in ecological indicators. The different maps indicate, first, that a link can be measured between practices and landscape dynamics and second, that the link does not correspond only to the type of production or plant cover a field carries but also to specific family/farm systems.

Selecting a few farms as examples, we can sum up how they relate to the patterns' diversities discussed above.

The Impacts of the Farmers' Whole-Systems

Different Family/Farm Systems

We noticed clear contrasts between the land farmed by families belonging to different cultural contexts. It was the case for the FO family farm, the MA farm adjacent to it, and the RC farm (see Table 2 and Figure 2). These three contrasting farms were therefore compared in terms of inputs and the use by birds of the landscape.

Farmer FO produces veal meat from a medium-sized semitraditional farm still divided into a certain number of fields. The farmer is from the area, where his father used to work as a sharecropper, and his spouse's parents were farmers in an adjacent village. The farm is not entirely located in Saint-André, where the initially farmed parts and house of the family are situated: all the land is rented from the members of the family, including relatives in the adjacent village. Still, the farmer's involvement in the village council remains in Saint-André. A lot of farming help is brought by the farmer's and his wife's parents (two are living with them and their children). Most of the household food, all the timber, and all of the food for the cattle come from the farm. The veal produce is sold through a small cooperative association with the benefit of a certificate (mother-fed produce). Using the CUMA cooperative organization necessitates little bought machinery. Although intensification is limited, the adaptation to family/farm conditions is of interest, but it needs the final quality of the produce to be socially valued.

The second group of fields is farmed by the MA family, and represents about one-half of a large farm located in another adjacent village, where the family lives in a very large house known as a castle. The MA family has migrated fairly

Table 2 Ecological Characteristics of the Land Farmed by the Three Farmers

		FO farm	MA farm	RC farm
TOT Total number of birds	Average	23.4	13.5	21.8
	Variance	44.7	38.4	22.4
	Median	22.7	12.4	21.3
NESP:	Average	10.0	5.8	9.8
Species richness	Variance	6.0	5.5	4.1
	Median	10.1	5.5	9.7
F1 AXIS SCORE	Average	0.1	0.7	0.4
Open landscape	Variance	0.3	0.4	0.2
	Median	0.0	0.7	0.5

Total number of birds: the total number of birds recorded singing or flying during a point count of 20 mn on a land unit of 6.25 na (grid of 250 m) during the nesting period. A male robin singing counts for two individuals (male + female).
Species richness: the number of different bird species recorded with the above-mentioned method.
F1 axis score: the score of the different pixels of 6.25 ha found on the farm territory on the first axis of the correspondence analysis (CA) modeling of bird community patterns. The score of the pixels depends on the list and abundance of the bird species recorded with the method (see Balent and Courtiade [1991] for details). This scores the ecological characteristics of the pixel on a gradient from woody landscape, then hedgerow network landscape, to open landscape.
The values of the three attributes are expressed for land units of 6.25 ha, i.e., the pixel of the bird community study (cells of 250 m/side). For example, a mean number of bird species of 10 on a 62.5-ha farm (10 pixels of 6.25 ha), is the average value of the number of bird species in the 10 pixels.

recently from a suburban area in the northeast of France, after they had been obliged to leave their previous farm due to the construction of a car factory. They have bought all the land they use on a GAEC basis (Groupement Agricole d'Exploitation en Commun = Agricultural Common Farming Association; the association is in this case between the father and two sons). The two brothers mainly ensure the farm work, helped by the wife of one of them who mostly does the accounting. The two parents do a little extra work and gardening. Although the whole family previously produced milk, sold directly to consumers, the brothers started new productions when they arrived. They raise cattle for meat and a few sheep, and some of the land is devoted to intensive growing of corn, wheat, barley, or soybeans. The brothers organized their own irrigation system, though not on the Saint-André territory concerned in this study. They are not members of the CUMA organization and have bought all of their machinery, which at first gave them an "individualistic" reputation. On the other hand, they have started work exchanges with other farmers, and are considered as "good farmers". Despite the crop and forage intensification, the labor resource and work organization inside the family is compatible with a certain amount of self-grown and raised food which maintains the way of life they consider good; almost all the cattle feeds come from the farm.

The RC farm, which is of medium size, is managed by a son from the village, where the whole farm is situated. The farm produces milk and has been intensified; the initially owned land has been expanded through renting from its

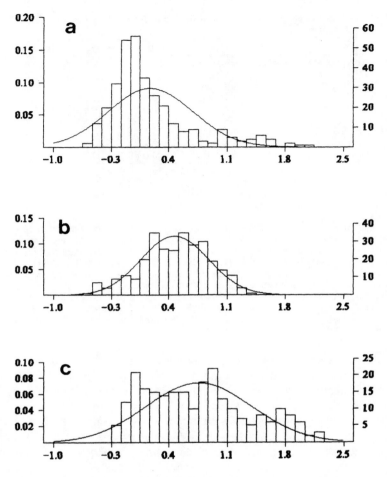

Figure 2. Distribution of the territory farmed by each family along the forest to cornfield gradient, defined from the changes in the bird community composition with correspondence analysis. (a) FO family farm; (b) RC family farm; and (c) MA family farm. To allow comparisons between farm data and ecological data from bird communities, we have split the pixel used for the bird community study (250 × 250 m) into 16 pixels of 62.5 × 62.5 m). The pixel area used to map the farm land is therefore, close to 0.4 ha.

neighbors. All the pastures and intensive meadows are devoted to animal feeding; so are almost all the cereal crops, and silage takes an important part in the forage system. In contrast to the intensive fields situated near the river Nere and southeast of the farm, some half-fallow slopes recently added to the surface are being used for the young cows. The managing of these complementary areas does not satisfy all of the cows' requirements, and a large amount of processed feed is bought from outside the farm. The crops, corn and grass silage, and the 45 cows

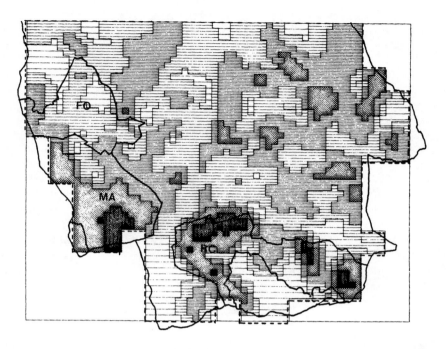

Figure 3. Distribution of nitrogen fertilizer inputs (increasing values from clear to dark).

require a considerable amount of work, which is done by the farmer and his wife. The other enterprises are limited to fattening a few ducks for production of *foie gras* for the family and for the regional market. The farmer's mother, who lives with the couple and their two children, is not able to help very much so the intensity of work is extreme: the farmer and his wife are extremely busy and there is no time margin which could be devoted to producing for self-consumption. The farm relies on the CUMA for a lot of machinery, which has allowed intensification without threatening the financial basis too much; but it is also dependent on external feed and an important amount of fertilizer input.

Levels of Inputs and Sizes of Fields

Among the inside characteristics of the practice systems as whole systems, we have isolated and emphasized some intensification signs which we believe to be related to the ecological state (see Figures 3 to 8). The maps show the marked contrasts which exist between our three examples. Noticeable contrasts also exist between the energy budgets of the three farms. We already know that other features are clearly related with and partly explain the ecological state: the size of the fields and the design of the hedges on the different farms. We must therefore take these into consideration when comparing the farms. The different patterns result in more or less hedges and trees, more or less cared for. We can

124 LANDSCAPE ECOLOGY and AGROECOSYSTEMS

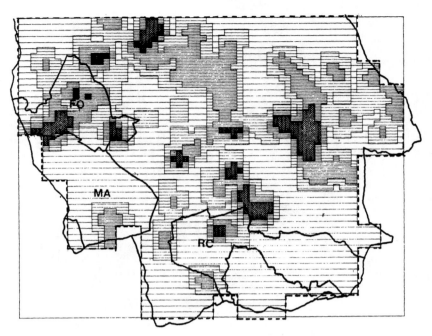

Figure 4. Distribution of organic manure inputs (increasing values from clear to dark).

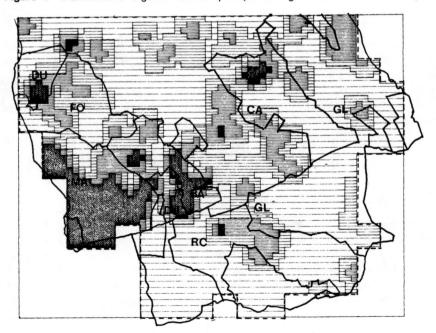

Figure 5. Distribution of herbicide inputs (increasing values from clear to dark).

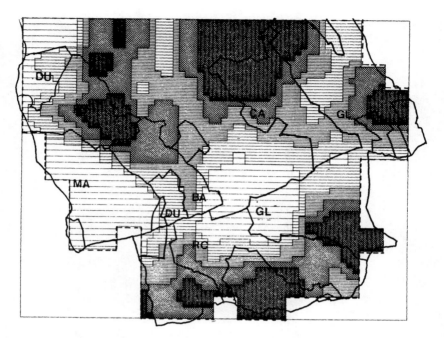

Figure 6. Total number of birds (increasing values from clear to dark).

Figure 7. F1: open landscape character (increasing values from clear to dark).

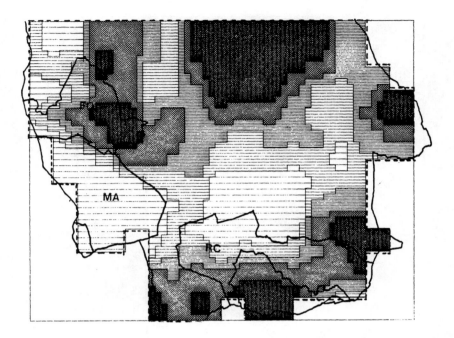

Figure 8. Species richness (increasing values from clear to dark).

see here the opposition between the lattice, still dense in the FO farm, which corresponds to all other aspects of FO's farming style, and the much larger, fewer, and open fields of the MA farm. The RC farm appears, in this as in other aspects, to be in an intermediate situation, probably corresponding to the various constraints of his forage-system during an evolving phase, when contrasted features coexist. For instance, some recently plowed fields have a very "open" structure, while some newly rented grazing areas were starting to evolve toward fallow land.

Comparison of the Results

If we consider the results based on the frequency histograms, they sum up and compare the quantitative evaluation of the landscape in relation to the areas belonging to the three farms, and therefore possibly the impact of the farmers' practices (see Table 2).

The strongest contrast for the three indicators is between the land farmed by FO and by MA. It corresponds to differences in their practices and their levels of intensification. The number of species and total number of birds are most significant in the FO land and least frequent (nearly by half) on the MA land, whose system requires very large fields and a much higher chemical input intensity. On the contrary, the MA territory has a very high position in relation with the open landscape ordination (F1 axis) while the FO territory is related to it in an opposite way.

As far as the open character of the landscape is concerned, the RC farm is also intermediate, which is also the case, as already stated, for the level of inputs. Nevertheless, compared to its position for the practices situation, the RC farm stands closer to the ecological performances of the FO farm as opposed to the MA farm. In the case of the total number of birds and the diversity of species, these results suggest that the explanation for this shift cannot be found among the inner characteristics of the farm. We must remember that the data concerning a group of fields can be influenced both by facts taking place at the scale of its direct vertical levels of control, such as the farm level, and by events occuring in adjacent or neighboring fields organized in a horizontal scale of control. (Hart, 1984; Allen and Starr, 1982).

What takes place, in this case, on the land adjacent to the farm? It appears that although the FO and MA farms are situated in more diversified spatial environments, many landscape units around the RC farm are covered with woodland, which has a noticeable impact on its ecological state. Residual woods inside a farmed landscape are therefore important for the conservation of agroecosystem ecological quality. The occurrence of woodland next to a given farm may be due to decisions in an adjacent farm or property and may also be controlled at the village level.

The next stage in the project is to follow the coevolution of the farms and the landscape. Since the present survey, important changes concerning the use of the fields have been decided at the village level, involving consolidation, restructuring of plots, and irrigation works. Therefore, we have planned to schedule our next farm survey after these changes in order to determine their consequences at the farm and parcel levels, for a comparison with a new landscape survey. We shall also consider the evolution of the woods areas and the hedgerow network in the whole village.

ACKNOWLEDGMENT

This study was made possible with a grant from the EGPN Committee of the Ministry of the Environment.

REFERENCES

Altieri, M. A. (1989). Agro-ecology a new research and development paradigm for world agriculture. *Agric. Ecosyst. Environ.*, 27:37–46.

Allen, T. F. and Starr, Th. B. (1982). *Hierarchy, Perspectives For Ecological Complexity.* The University of Chicago Press. Chicago, IL.

Balent, G. and Courtiade, B. (1992). Modelling bird communities/landscape patterns relationships in a rural area of South Western France. *Landscape Ecol.*, (6) 3:195–211.
Balent, G. (1987). Structure Fonctionnement et Évolution d'un Système Pastoral. Thèse de Doct. d'Etat. Université de Rennes I.
Balent, G., Baudry, J. and Sauget, N. (1989). Apport des modèles écologiques à l'analyse des modifications de l'activité agricole au niveau d'une petite région. *Sadoscope*, 44:65–70.
Baudry, J., Burel, F. and Balent, G. (1988). Ecological consequences of changes in agricultural systems: some methodological aspects and cases study in France. *Int. Ecol., (Intecol Bull.)*, 16:35–46.
Baudry, J. (1989). Interactions between agricultural and ecological systems at the landscape level. *Agric. Ecosyst. Environ.* 27:119–131.
Braudel, F. (1986). *L'Identité de la France*. Arthaud.
Buttel, F. H. and Larson, O. W. (1979). Farm size, structure and energy intensity: An ecological analysis of U.S. agriculture. *Rural Sociol.*, 44:471–488.
Buttel, F. H. (1980). Agriculture, environment and social change: some emergent issues. In: FH. Buttel and H. Newby (Editors). *The Rural Sociology of the Advanced Societies*, F. H. Buttel and H. Newby (Eds.), Allanheld, Osmun & Co., Lanham, MD, 453–488.
Coleman, D. C. (1989). Ecology, agroecosystems, and sustainable agriculture. *Ecology*, 6,70:15–17.
Daly, H. E. (1990). Toward some operational principles of sustainable development. *Ecol. Econ.*, 2, 1:1–6.
Deffontaines, J. P. (1985). Etude de l'activité agricole et analyse du paysage. *L'Espace Géogr.*, 1:37–47.
Deleage, J. P., Julien, N., Sauget, N. and Souchon, C. (1979). Eco-energetics analysis of agricultural systems: the French Case. *Agro-ecosystems*, 5:345–365.
Deleage, J. P., Sauget, N. and Souchon, C. (1978). Economie-Ecologie, l'analyse énergétique des écosystèmes ruraux. Commun. Colloque SFER Ecologie et Société. *Economie Rurale*, 5.
Eizner, N. (1985). *Les Paradoxes de l'Agriculture Française*. L'harmattan, Paris.
Forman, R. T. T. and Baudry, J. (1984). Hedgerows and hedgerow networks. *Landscape Ecol.*, 8:495–510.
Godelier, M. (1978). Reproduction des Ecosystémes et Systèmes sociaux. Commun. Colloque SFER Ecologie et Société. *Economie Rurale*, 5.
Godelier, M. (1984). *l'Idéal et le Matériel*. Fayard, Paris.
Golley, F. B. and Golley, P. (1988). Environmental consequences of changing agricultural policy and practice. Ecology International, 15 (Intecol Bulletin).
Golley, F. (1987). Introducing landscape ecology. *Landscape Ecol.*, 1:1–3.
Golley, F. B. (1987). Future directions in landscape ecology research. *Landscape Ecol.*, 4:191–192.
Gulinck, H. (1986). Landscape ecological aspects of agroecosystems. *Agric. Ecosyst. Environ.*, 16:79–86.
Hart, R. D. 1984). Agro-ecosystem determinants, in: *Agricultural Ecosystems, Unifying Concepts,* Lowrance, R., Stinner, R. and Garfield, (Eds.), John Wiley & Sons, New York, 105–118.
Hutter, W. (1978). Bilan Énergétique des Cultures. 2nd Congr. Eur. Écon. Agric.,4–9.

Jollivet, M. (1974). L'analyse fonctionnelle structurelle. *Les Collectivités Rurales Françaises*, T.2., Armand Colin, 155–229.

Jollivet, M. (1984). Réflexions Méthodologiques sur le Programme Causses-Cévennes. Rapport 81–82.

Jollivet, M. (Ed.) (1988). *Pour une Agriculture Diversifiée*. L'Harmattan, Paris.

Lemonnier, P. (1983). L'étude des Systèmes Techniques, une urgence en technologie culturelle. Tech. Cult., 1:11–25.

Mathieu, N. and Dubosq, P. (1985). Voyage en France par les pays de faible densité. Ed. CNRS.

Nassauer, J. I. and Westmacott, R. (1987). Progressiveness among farmers as a factor in heterogeneity of farmed landscapes. *Landscape Heterogeneity and Disturbances Ecological Studies*, M. Goigel Turner, (Ed.), Springer-Verlag, New York, 65.

Norgaard, R. B. (1986). The epistemological basis of agroecology. *Agroecology: the Scientific Basis of Alternative Agriculture*, M. A. Altieri (Ed.), Westview Press.

Osty, P. L. (1978). L'exploitation vue comme un système. *BTI*, 326:43–49.

Osty, P. L. (1988). Un essai pour décrire les élevages en termes de système technique. *Eleveur Troupeau et Espace Fourrager*, Gibon et al. (Eds.), Etudes et Recherches sur les Systèmes Agraires et le Développement, 11–15.

Paul, E. A. and Robertson, G. P. (1989). Ecology and the agricultural sciences: a false dichotomy? *Ecology*, 70:1594–1596.

Perget, S. (1983). Approches des Systèmes d'élevage bovin des coteaux de Gasgogne. Mémoire de fin détudes. ENSAA Dijon, INRA-SAD, Toulouse.

Pernet, F. (1982). *Résistances Paysannes*, Presses Universitaires de Grenoble.

Pimentel, D. and Pimentel, M. (1979). *Food, Energy and Society*, Edward Arnold, London.

Pimentel, D., Wen Dazhong and Giampietro, M. (1990). Technological changes in energy use in U.S. agricultural production. *Agroecology*, S. Gliessman (Ed.), Springer-Verlag, New York, 305–321.

Prouzet, F. (1983). Systèmes fourragers en zones de coteaux. Mémoire de fin détudes, ESA Purpan et INRA, Toulouse.

de Ravignan, F. (1981). Transects paysagers et reconnaissance brève du canton d'Aurignac (Haute-Garonne), INRA-SAD, Toulouse.

Remondière, S. 1982. Analyse de la situation foncière des exploitations agricoles sur le canton d'Aurignac. Mémoire de fin d'étude, ESA Purpan, INRA-SAD, Toulouse.

Remy, J. (1982). Le métier d'agriculteur. Façons de produire et façons d'être des agriculteurs Sarthois, INRA, Paris.

Sauget, N. (1979). Energie et Systèmes de Production, analyse de 4 exploitations-types. CEGERNA et Laboratoire d'Economie générale et appliquée, de Paris VII, Paris.

Sauget, N. and Maresca, B. (1986). La diversité des façons de produire des agriculteurs. Commun. Colloque DMDR, Paris, avril 1986.

Umbach, E. (1989). Socio-economic factors as causal factors in the ecosystems. *Ecol. Model.*, 46:305–308.

Zonneveld, I. S. (1989). The land unit: a fundamental concept in landscape ecology and its applications. *Landscape Ecol.*, 3:67–86.

12 Role of Natural Vegetation in the Agricultural Landscape for Biological Conservation in Sicily

L. DI BENEDETTO
F. LUCIANI
G. MAUGERI
E. POLI MARCHESE
S. RAZZARA

Abstract: *Sicily has been subjected to heavy human impact, as has much of the Mediterranean region, and therefore has a vegetative landscape dominated by agriculture. The agricultural landscape, often characterized by monocultures, has few elements of natural vegetation. The landscape is therefore monotonous, with limited aesthetic appeal and low ecological diversity. Intervention is needed to increase diversity, favor conservation of wildlife, and to lead to the extension and recovery of natural vegetation as well as to an improvement in the balance within the whole region. The integration between natural vegetation and agroecosystems can produce a harmonious landscape without uniformity and therefore with a distinctive character.*

INTRODUCTION

Sicily, like most of the Mediterranean region (Naveh and Dan, 1973), has been subjected to antropic intervention for thousands of years. The constant action of man has been exerted through grazing, wood cutting, fire, and especially crop production (Ramade, 1978).

Human intervention has caused radical alterations in the original vegetation of Sicily. Instead of deciduous woodlands on the Mediterranean mountain and supra-Mediterranean belts (*Querco-Fagetea* communities), there are now wide areas of herbaceous vegetation (*Arrhenatheretea* communities) used mainly for grazing, with fragments of degraded woodland vegetation. The evergreen woodlands on the lowland Mediterranean belt (*Quercetea ilicis* communities) have now been substituted to a great extent by crops (Poli Marchese, 1982).

The lack of diversity in the vegetation and the dominance of crops over extensive areas has led to an extremely monotonous landscape. Apart from aesthetic disadvantages, there are also negative biological and ecological consequences.

EFFECTS OF HUMAN INTERVENTION

Cultivated fields in Sicily are very widespread and they reach quite high altitudes due to the mildness of the climate. In the Cesar area, for example, wheat is cultivated up to an altitude of 1500 m (Jannaccone, 1950); apple orchards on Mount Etna go up to about 1500 m, and hazel groves in the Messina district reach an altitude of 1000 m.

Citrus orchards are the main arboreal cultivations in the warmer zones wherever water is readily available; elsewhere, vineyards and other typical Mediterranean cultivations (olive and almond groves) are widespread.

The dominant crop in Sicily, however, is wheat, which is grown at sea level and reaches to the supra-Mediterranean belt, thus occupying about 20% of the island and 30% of its agricultural area.

Progressive destruction of natural vegetation has increasingly reduced the woodland area. In the 1950s this area was small, being only about 4% of the whole, whereas the whole island territory, excluding the summit zone of Mount Etna (>2000 m above sea level) is potentially woodland. Apart from woodlands, the natural vegetation now consists mainly of degraded herbaceous and shrub vegetation localized in areas which are not used for agricultural purposes. Overall, the vegetation is therefore dominated by crops. Frequently there is a marked dominance of a monoculture, with 70% of the total agricultural area being taken up by wheat (31%), olive (20%), grapevine (13%), and citrus (6%) (Paolino, 1986).

Progressive abandonment of farming over the last few decades has not helped to improve the environmental conditions of the territory, already compromised by antropic intervention, nor improved its aesthetic appearance. The abandonment of farming, mainly limited to areas less suitable for agricultural use due to soil or geological factors, has in many cases favored soil erosion. The vegetation which forms in these areas consists of nitrophilous-ruderal communities (*Brometalia rubenti-tectori* communities) and degraded herbaceous communities (*Thero-Brachypodietea* and *Helianthemetea guttati* communities) that have little value from either an economic or a landscape point of view. These are mainly poor grazing areas with examples of the most involute stages of vegetation (Poli Marchese et al., 1988). The most widespread types of vegetation in the island include garigue of *Thymus capitatus* on calcareous soil, the heaths of *Cistus* sp.pl. in areas characterized by frequent fires, the *Lygeum spartum* community on argillaceous soil, and *Ampelodesmos mauritanicus, Cymbopogon hirtus,* and *Asphodelus microcarpus* communities on the poorest soils.

The progressively increasing human intervention is continuing to have negative effects on the natural vegetation of Sicily. The increase in the intensity of cultivation has favored an increment of allochthonous species, such as archeo-

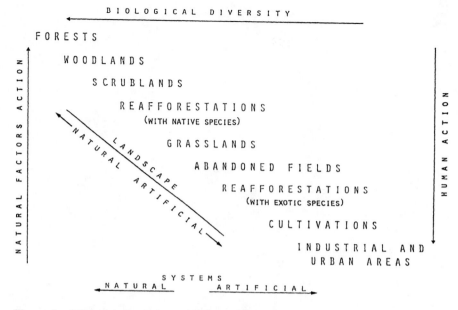

Figure 1. Natural and artificial ecosystems in relation to human intervention.

phyte, neophyte, and tropical species adapted to environments subject to human intervention such as *Fumaria officinalis, Veronica persica, Papaver rhoeas, Oxalis pes-caprae, Amaranthus* sp.pl., and *Cyperus rotundus*. Human intervention has at the same time favored the spread of some autochthonous species such as *Stellaria neglecta* (Maugeri, 1979).

The drastic reduction of areas once occupied by natural vegetation has jeopardized the survival of elements of autochthonous flora. This flora, which once occupied the whole area of the island, is now limited to small, fragmentary areas (about 15% of the total range) and is often strongly subjected to human activity.

The woodland formations of the past have now become more or less degraded, isolated fragments localized at the highest altitudes, on the edges of crop fields, and in areas which cannot be exploited for agricultural purposes such as rock formations, inaccessible areas, etc.

Such relicts of natural vegetation are a refuge for plant and animal species which have disappeared elsewhere. Human activity has therefore led to the disappearance of natural ecosystems and to a modification of the individual character of the island (Figure 1).

CRITERIA FOR INTERVENTION

In spite of the present situation in Sicily, we are convinced that certain operations could be undertaken to ensure survival of the individual animal and plant species and the relative coenoses, and also of entire ecological systems.

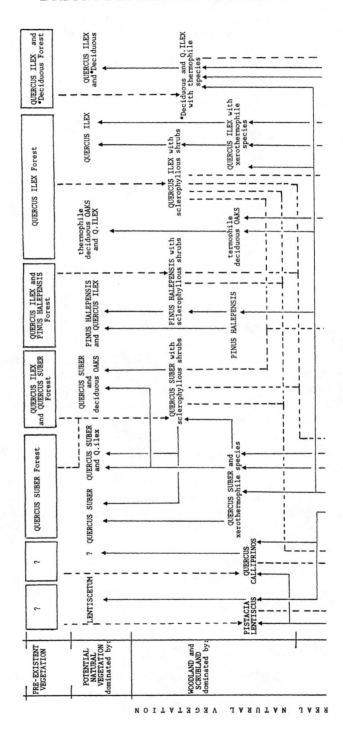

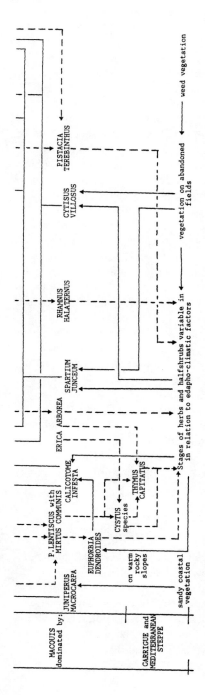

Figure 2. Successional pathways of Mediterranean evergreen vegetation in Sicily; → progressive successions; --→ regressive successions. By the term "deciduous" we mean such species as: *Fraxinus ornus, Ostrya carpinifolia, Acer obstusatum,* and *Ulmus campestris.*

Some of the ecological principles which are involved in order to achieve a rebalance of the environmental situation are outlined below. For this purpose, an important element is the vegetation which is the essential component of the landscape. Intervention in the composition of vegetation could therefore be a starting point for redressing the balance. Intervention should, where possible, aim at restoring the original balance of species forming the natural vegetation. The actual potential vegetation should therefore be borne in mind (Tüxen, 1956).

Investigations so far carried out on the vegetation in Sicily have provided information with regard to the various altitude belts and their plant communities, not only from the floristic and ecological points of view but also in terms of dynamic relationships. In particular, studies on vegetation dynamics have made it possible to fix the main stages in the dynamic of the vegetation, i.e., both progressive and regressive successions. Through detection of progressive successions, it has been possible to define dynamic series and thus determine the actual potential vegetation. If one takes as an example the belt of evergree Mediterranean vegetation in Sicily (Gentile, 1968), one finds that numerous series can be distinguished. This is synthesized in the diagram (Figure 2) reported by Poli Marchese et al. (1988). Knowledge of the various dynamic levels of the vegetation and definition of the site occupied by each plant community in the series or in the dynamic series to which it belongs is essential if action is to be undertaken to enhance, accelerate, or restore natural processes in order to ensure recovery of the potentiality of degraded soils.

In particular, in areas not directly utilized for agriculture, measures should be taken to enhance preservation and formation of natural vegetation and to recover degraded environments such as abandoned fields, impoverished grazing lands, degraded aspects of the natural vegetation, areas along the edges of crop fields, and the banks of ditches (Poli Marchese et al., 1990).

These interventions, when carried out with adequate ecological territorial planning, determine soil preservation, prevent further erosion, and lead to the recovery of productivity. This would allow the maintenance and increase of autochthonous species and plant communities as well as enhancing biological diversity, thus offering a better guarantee for the maintenance of a harmonious landscape (Naveh, 1971; 1975).

CONCLUSIONS

The objective of an integrated landscape can be achieved by adequate legislation aimed at protecting the environment, without preventing rational exploitation of resources. In recent years increased awareness of such problems has led to the establishment of natural parks and reserves. These areas are significant not only with regard to conservation of the natural biological potentialities, but also in that they represent valid reference points for territorial intervention for the purpose of increasing biological diversity and productivity.

Actions are required to advise the government to increase such areas in Sicily in order to take into account knowledge of the ecological factors (Tomaselli, 1971a, 1971b). The principles of correct environmental management currently apply only in natural parks and protected areas in general, and therefore need to be extended to the whole island. If adequate environmental management is achieved over the whole territory of Sicily, the conservation, recovery, and spread of natural vegetation will have an important role in the ecology of the landscape. It will ensure biological diversity and the conservation and diffusion of animal and plant species and the relative coenoses (Poli Marchese, 1990).

The integration of natural vegetation with agroecosystems will not only favor the maintenance of natural resources, but it will also create a less monotonous, aesthetically more harmonious landscape with characteristics of its own (Giacomini, 1973).

REFERENCES

Giacomini, V. (1973). Quanto si potrà resistere seguitando su questa linea? Agricoltura e difesa dell'ambiente. 1st. Tecnica e Propag. Agraria, 25–31.

Gentile, S. (1968). Memoria illustrativa della carta della vegetazione naturale potenziale della Sicilia. *1st Bot. Lab. Critt. Univ. Pavia*, 40:1–114.

Jannaccone, A. (1950). *Coltivazioni Erbacee (Cereali)*. Casa del Libro (Ed.), Catania, 586.

Maugeri, G. (1979). La vegetazione antropogena della Sicilia. Primo quadro sintetico. *Boll. Acc. Gioenia Sci. Nat. Catania*, 13, 13(10)(4):135–159.

Naveh, Z. (1971). The conservation of ecological diversity of Mediterranean ecosystems through ecological management. *The Scientific Management of Animal and Plant Communities For Conservation*, E. Duffy and A. S. Watt (Eds.). Proc. 11th Symp. Br. Ecol. Soc., Norwich 1970. Blackwell, London, 605–622.

Naveh, Z. (1975). Degradation and rehabilitation of Mediterranean landscapes. Neotechnological degradation of Mediterranean landscapes and their restoration with drought resistant plants. *Landscape Plann.*, 2:133–146.

Naveh, Z. and Dan, J. (1973). The human degradation of Mediterranean landscapes in Israel. Mediterranean type ecosystems. *Ecological Studies*, F. di Castri and H. A. Mooney (Eds.), Springer-Verlag, New York, 7:373–390.

Paolino, G. (1986). *L'Agricoltura Italiana in Cifre*, La Nuovagrafica. Catania, 290.

Poli Marchese, E. (1982). Antropizzazione del paesaggio vegetale nella regione Mediterranea. *Atti Acc. Medit. Sci. Catania, Vol. 1.*, 1:45–58.

Poli Marchese, E. (1990). La vegetazione nella caratterizzazione ecologica del paesaggio. *Ecologia del Paesaggio, Prospettive Teoriche e Pratiche in Italia*. Semin. Naz. Studi SITE-IALE, Italia, Parma 25.5.1990, 50.

Poli Marchese, E., Di Benedetto, L. and Maugeri, G. (1988). Successional pathways of Mediterranean evergreen vegetation on Sicily. *Vegetatio*, 77:185–191.

Poli Marchese, E., Maugeri, G. and Di Benedetto, L. (1990). Studio della vegetazione per il recupero alla produttivita di aree marginali della Sicilia. Sistemi agricoli marginali Sicilia. C.N.R.-Progetto finalizzato I.P.R.A., Palermo, 41–65.

Ramade, F. (1978). Eléments d'écologie appliquée. Action de l'Homme Sur la Biosphère. McGraw-Hill, New York, 576.
Tomaselli, R. (1971a). La vegetazione nell'ambito della pianificazione ecologica territoriale. *Problemi dell'Ecologia,* Aziende Bardi, Rome, 119–198.
Tomaselli, R. (1971b). Il Significato della Vegetazione Naturale Potenziale Nella Pianificazione Territoriale Ecologica, XIII Congr. Sci. Biol. Morali, Ottawa.
Tüxen, R. (1956). Die heutige potentielle natürliche Vegetation als Gesenstand der Vegetations-kartierung. Angew. Pflanzensoz., Stolzenau/Weser, 13:7–32.

PART II
Agroecosystems

13 Patterns of Change in the Agrarian Landscape in an Area of the Cantabrian Mountains (Spain) — Assessment by Transition Probabilities

A. GÓMEZ-SAL
J. ALVAREZ
M. A. MUÑOZ-YANGUAS
S. REBOLLO

Abstract: *A systematic sampling on aerial photos was used to distinguish 17 types of "land use" with regard to vegetation physiognomy in an area of 35 km² in the Cantabrian mountains (Spain). The sampling was carried out on three different series of aerial photographs (taken in 1957, 1974, and 1985) which showed the change from a traditional agricultural situation to the present day, characterized by a drastic reduction of at least 50% in the population. The analysis of the land use change by means of Mutual Information and transition probabilities (tested by the Monte Carlo method) allows the evaluation of the degree of landscape transformation. In the first period (1957 to 1974), the change affected 35% of the area. There was a loss of crop area (8% of the whole land) that was converted into different types of grazing or scrubwood lands, whereas some ancient pastures were replaced by pine plantations. In this period the analysis revealed an ordered sequence of changes, with many transitions between uses which are close from a successional point of view. During the second period, which showed a change in 38% of the area, there was a transition from shrub pasture to more complex woody stages with trees (19%) and the virtual disappearance of crops. The crops were replaced by pastures and meadows. In the second period, jumps between uses which are further apart in the succession were generally more predominant. In general, there was a simplification of agricultural and livestock uses and a corresponding increase in forest and scrub.*

INTRODUCTION

About 70% of Spain is defined as mountainous and 38% is considered to be within the zone of mountain agriculture according to European criteria. With few exceptions, the whole area has been affected by a process of population loss which has taken place since the 1950s and has been especially marked in the continental mountain ranges with adverse climates. These regions (southern slopes of the Cantabrian mountains, central Pyrenees, Iberian system, and parts of the Central system) maintained their highest population during the first decades of this century, as a direct result of a process which took place in the middle of the last century (Chocarro et al., 1990; García-Ruíz and Lasanta, 1990). The economy was self-sufficient and by means of specialized exchanges with other areas (Galindo, 1987; Gómez-Sal, 1988) a heterogeneous and multiagricultural landscape was maintained, supported by a complex social organization with strict rules as to the management of resources (Montserrat, 1977; González Bernáldez, 1982, 1983). The necessity of feeding a growing population meant that almost all available land was plowed and planted with unsuitable cereal crops even in inaccessible places (Puigdefábregas, 1987).

The use of the summer grazing land (of high productivity, but available for only a few months, between June to September) represented an important resource for the local population, because it was impossible to feed a large number of animals in winter with their own resources. In the majority of villages in the high Cantabrian mountains, the exploitation of the pastures was based on complex transhumance systems, which brought together two areas of complementary production, i.e., summer grazing land in the northern mountains, and the savannah-like forest (called "dehesa" in Spanish) of holm oak (*Quercus rotundifolia* Lamk.) in the southern lowlands, sometimes as much as 600 or 800 km apart.

At present, although the use of summer grazing areas has been maintained despite great problems, the multiagricultural agrarian system has suffered a major transformation in the mountains (Gómez-Sal, 1988). This generally tends towards a specialization in livestock breeding, moving from crops to pastures, and the substitution of varied livestock (goats, sheep, and horses) by semi-extensive cattle raising which require less manpower. This kind of management is clearly not sufficient to maintain the whole landscape in production. The abandonment of the land, with the subsequent invasion of shrubs, is the most common feature in the main part of the region. The population becomes restricted to areas with employment opportunities in the tertiary sector (i.e., tourism in the Pyrenees or Picos de Europa) or where there is still some industrial exploitation (mining in the Cantabrian mountains — León and Palencia provinces).

In this study the patterns of change in the landscape are highlighted by an analysis on the changes in land use which have taken place at three times during the 1957–1985 period, using aerial photographs.

STUDY AREA

The study was carried out in a mountain valley of about 35 km^2 in the Cantabrian mountains in northern Spain (see Figure 1a). The complex geological structure is made up of various paleozoic materials and the altitude ranges from 1050 to 1800 m. The woodland cover consists mainly of oak forest (*Quercus*

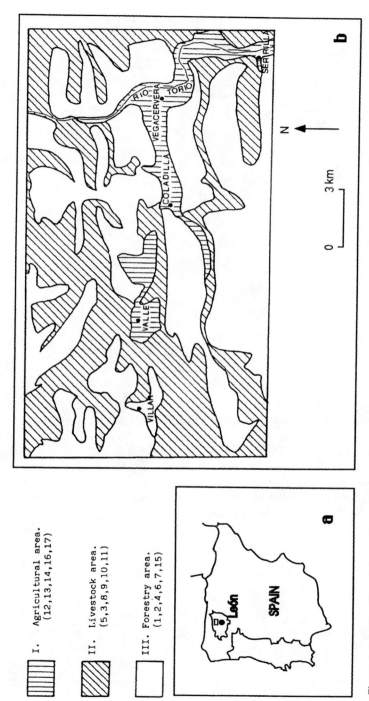

Figure 1. (a) Location of the study area, and (b) zones which can be distinguished by their predominant use: agricultural (I), livestock raising (II), and forestry (III).

Table 1 Types of Land Use Considered in Study Area

1.		Rock
2.		Forest
3.		Open Forest
4.		Trees (40–75%) & Scrub (25–60%)
5.		Trees (<40%) & Scrub and Grassland (>20%)
6.		Scrub (>75%)
7.		Scrub & Trees (<40%)
8.		Scrub (>20%) and Grassland
9.		Scrub (>20%), Grassland and Rock
10.		Limestone pasture
11.		Grassland and Scrub (<20%)
12.		Meadows
13.		Crops
14.		Villages
15.		Forest restoration
16.		Road
17.		River

Differentiation was widely based on a vegetation physiognomy criteria.

pyrenaica Willd). The most common shrubs are *Genista florida* L., *Cytisus scoparius* (L.) Link, *Cytisus purgans* (L.) Boiss., *Erica arborea* L., and *Rosa* spp. In the grasslands, perennial species are predominant. During the period analyzed, a transition took place from a self-sufficient agricultural system with diversified production, to an almost industrialized livestock-raising monoculture. The population diminished by more than 50%.

METHODS

Sampling

We selected 485 sampling points distributed over a 250-m-wide grid. Sampling was carried out on three series of aerial photographs corresponding to the years 1957 (t_1), 1974 (t_2), and 1985 (t_3). Table 1 identifies 17 types of land use, the definition of which were largely based on vegetation physiognomy. For analysis, the whole region was divided into three large areas according to the prevalent production type at t_1, as shown in Figure 1b.

Analytical Procedures

The importance of the overall change in land use has been estimated by means of both Mutual Information (MI) (Abramson, 1963) and Diversity (Shannon-Weaver) Indexes applied to the coincidence matrix between uses of the periods considered (t_1, t_2, and t_3). The use of these methods in ecological cartography can be seen in De Pablo et al. (1987 and 1989).

In order to detect patterns of change, the calculation of transition probabilities (Gibson et al., 1983; Sltayer, 1977) has been carried out (conditioned probability matrices between uses of different years) on the four possible cases: pairs t_1–t_2, t_2–t_3, t_1–t_3, and triplets t_1–t_2–t_3. The statistical significance of the conditioned probabilities was analyzed by the Monte Carlo permutation test (Edington, 1969).

RESULTS AND DISCUSSION

Overall Changes

Using the MI index as a measure of the resemblance between the spatial distributions of uses in the years under comparison, differences of 65% were found in the intervals t_1–t_2 and t_2–t_3, and 53% in the total period t_1–t_3.

Changes in Dominance

As can be seen in Figure 2 (a and b) the changes in dominance are due to the fact that the area occupied by crops (use 13) and grazing (uses 8, 11, and to a lesser degree, 4 and 7) increased. However the area occupied by meadows (use 12) and that occupied by scrub and shrubs (uses 6, 5, 2, and to a lesser degree, 4 and 7) increased. In t_1 the most widespread uses were "pastures" and crops, whereas in t_3 scrubby woodland and meadows predominated. The original grazing zones had in t_3 a slight presence because of their invasion by shrubs. Cattle breeding is concentrated in the more productive plots.

Diversity Changes

Taking into account the entirety of the uses under consideration it can be said that diversity increases with time in the whole area (see Table 2). Analysis of the data according to the prevalent production type at t_1 (see Figure 1b) — agricultural (I), livestock raising (II), or forestry areas (III) — shows that the above-mentioned increase of diversity is due to the latter two areas. This is a result of the internal dynamics released by the scrub and woodland formations after the abandonment of ancient grazing lands. In the former area (agricultural), a reduction in diversity occurred because of the shift from crops to meadows. It can be seen (Figure 2c) that the area of land occupied by agricultural uses (I) diminishes, due to the abandonment of some plots formerly used for crops. The area used for livestock raising (II) is also reduced because of the abandonment of traditional grazing practices in the remoter areas. However, the area given

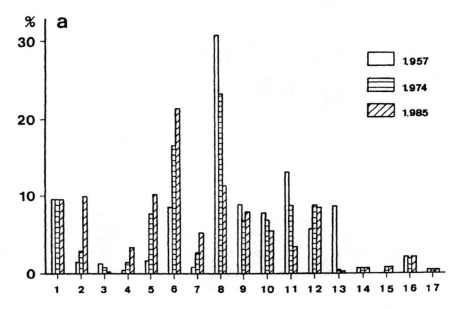

Figure 2. (a) Evolution of the relative area occupied by each land use type in the three periods considered. (b) Ranking of land uses according to the occupied area (in percentage) at each of the three periods studied. Arrows indicate change in relative position. (c) Evolution of the uses grouped according to predominant vocation: agricultural, livestock, and forestry.

over to forestry, use (III), increased because of colonization of the abandoned areas by woodland and scrub.

Patterns of Change

The significant relationships ($p < 0.01$) assessed (Monte Carlo method) between uses at the different time intervals studied are represented in Figures 3a, b, c, and d. As may be seen, most changes occur in the direction of the ecological succession. Forest and meadows behave as final, more stable, situations in the sequence of changes. Figure 3d emphasizes some t_2–t_3 transitions which depend on the initial t_1 situation.

Transition $t_1 \rightarrow t_2$

Crops (13) are replaced by meadows (12) and then remain as such. Pastures (8, 9, 10, and 11) tend to stay the same or to evolve into different types within the group ($9 \rightarrow 8$; $8 \rightarrow 10$). Scrub and woodland (5, 3, 7, and 2) develop into more mature and woodier states, and eventually into forests. In the t_1 period this process took place by means of progressive linear changes and by transitions between close situations. Finally, there is a series of uses which do not change.

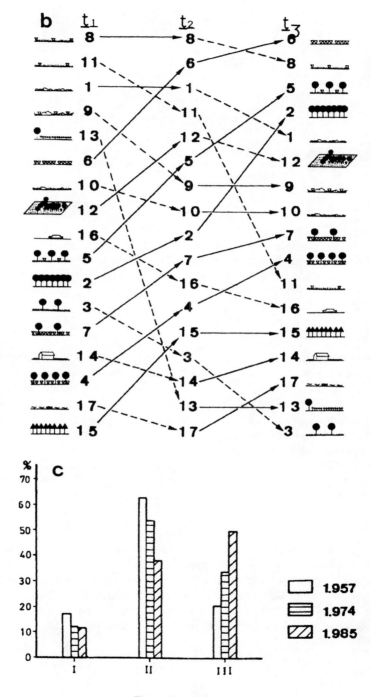

Figure 2 (continued).

Table 2 Diversity Values

	t_1	t_2	t_3
Whole area	3.20	3.38	3.47
I	1.69	1.31	1.27
II	1.99	2.19	2.24
III	1.61	1.93	2.13

Diversity values (Shannon index) calculated for the whole area and for each area according to its predominant use in t_1; (I) agricultural, (II) livestock, and (III) forestry areas. Numbers indicate evolution of the diversity in the three areas.

$t_2 \rightarrow t_3$

Pastures (8, 9, 10, and 11) show regression towards scrub (from $8 \rightarrow 9$ and $8 \rightarrow 6$) because of shrub invasion. The scrub itself evolves toward more mature woodland. Scrub and woodland gradually become more mature even though, on this occasion, they manage this by means of complex leaps rather than linear changes. In comparison with the former period, this second period (t_2–t_3) shows a more varied range of significant relationships.

$t_1 \rightarrow t_3$

The stony pastures (9 and 10) stay the same or develop slowly toward very close ''uses'' with a little more scrub and pasture. However, pastures with deeper soil and scrub (6 and 8) have a greater potential for development, and changed on abandonment into more mature states, with a tendency to the formation of forest. Meadows present in t_1 remain as such through to t_3 and increase from land that was originally occupied by crops. In spite of the scarcity of pasture and scrub woodland uses (3 and 7) in t_1, there was a trend shown toward forest. The uses indicative of infrastructure — 14 (buildings) and 16 (roads) — remained constant throughout the 30 years, and that also highlights the low investment or industrial development in the area.

$t_1 \rightarrow t_2 \rightarrow t_3$

All significant triplet relationships are displayed in Figure 3d, showing an excellent synthetic overview of the overall changes. The crops were mainly displaced by meadows, although some were abandoned and developed into scrub and woodland. Some pastures (9 and 10) stayed the same or developed in different directions. However, the other uses, 8 and 11, underwent significant changes towards scrub or woody scrub, eventually becoming forest. Some areas were also used for afforestation (use 15).

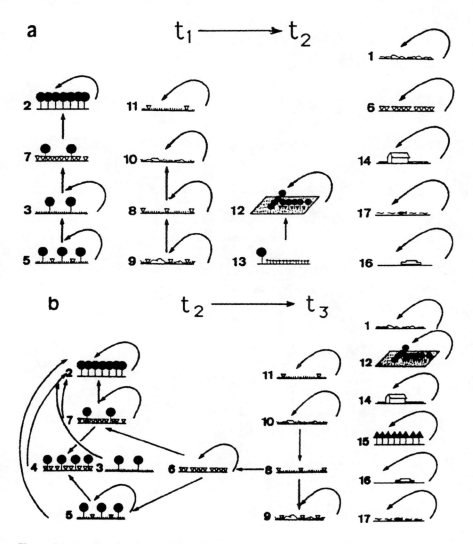

Figure 3. Land use changes which were statistically significant (*) according to the Monte Carlo test: (a) t_1–t_2 period, (b) t_2–t_3, (c) t_1–t_3, and (d) t_1–t_2–t_3.

CONCLUSION

The overall change in the agrarian landscape shows a general trend toward an increase in the wooded area — trees and scrubs — and a decrease in crops which is considered to be a successional trend. The results show different changes

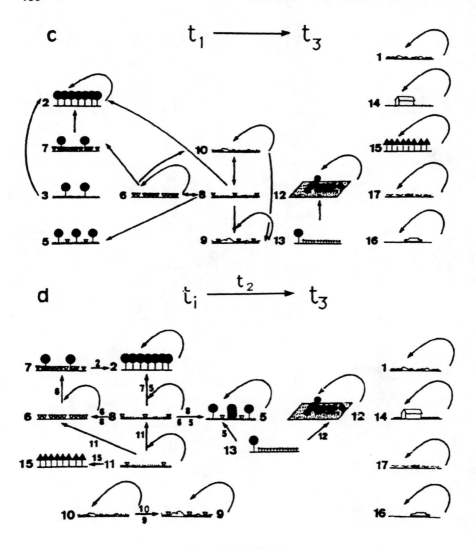

Figure 3 (continued).

in patterns in the two intervals (t_1–t_2, and t_2–t_3). In the first case, the analysis revealed an ordered sequence of changes, so that the transitions detected were between uses which are close in the succession. In the second, the situation is more complex. Jumps between uses which are further apart in the succession are generally more predominant. The first case would indicate an initial delicate balance in land uses and a state of searching, testing, and indecision in changing uses. The frequency of transitions between close stages (which suggests movement backward and forward) and lack of change, reveals that definite abandonment has not occurred yet. On the other hand, a certain degree of irreversibility or a more definite character in the changes seems to occur in the second period. There is no significant decrease in uses where pastures and meadows predominate, indicating that the present total number of livestock has been maintained in the areas where such uses are present. However, in areas dominated by scrub or forest, livestock have disappeared. A limitation of livestock enterprises — mainly dairy cows — in the more productive zones has also happened. Changes in the diversity indicate that the simplification of uses in the old agricultural area (I) is due to loss of crops. On the remaining land (II and III) an increase in the diversity of uses based on the different stages in succession from scrub and pastures to forest actually took place and involves different types of wooded vegetation.

ACKNOWLEDGMENTS

This research is a part of the project "Function of grazing ecosystems" (PB87-0451) supported by the DGICYT, (Spain).

REFERENCES

Abramson, N. (1983). *Information Theory and Coding*. McGraw Hill, New York, 218.
Bertrand, G. (1972). Ecologie d'un espace geographique: Les geosystems du valle de Prioro (Espagne du Nord-Ouest). *L'Espace Geographique*, 2:113–128.
Chocarro, C., Fanlo, R., Fillat, F. and Marín, P. (1990). Historial of natural resources use in the central Pyrenees of Spain. *Mount. Res. Dev.*, 10(3):257–265.
De Pablo, C. L., Gómez-Sal, A., and Pineda, F. D. (1987). Elaboration automatique d'une cartographie écologique et son évaluation avec des paramètres de la théorie de l'information. *L'Espace Géographique*, 2:115–128.
De Pablo, C. L., Martín de Agar, P., Gómez-Sal, A., and Pineda, F. D. (1988). Descriptive capacity and indicative value of territorial variables in ecological cartography, *Landscape Ecology*, 1(4):203–211.
Edington, E. S. (1969). Randomization test. *J. Psychol.*, 57:445–448.
Galindo, J. L. M. (1987). Poblamiento y actividad agraria tradicional en León. Estudios de geografia rural. Junta de Castilla y León, Valladolid. 167.
García-Ruíz, J. M. and Lasanta, T. (1990). Land-use changes in the Spanish Pyrenees. *Mount. Res. Dev.*, 10(3):267–279.
Gibson, C. W. D., Guilford, T. C., Hambler, C. and Sterling, P. H. (1983). Transition matrix models and succession after release from grazing on Aldaba atoll. *Vegetatio*, 52(3):141–153.
Gómez-Sal, A. (1988). Ecosistemas rurales. *Elementos Básicos Para la Educación Ambiental*, F. D. Pineda et al. (Eds.), Ayuntamiento de Madrid, Madrid, 53–77.
González Bernáldez, F. (1981). *Ecología y Paisaje*. Editorial Blume, Madrid, 195.
González Bernáldez, F. (1983). La preservación del paisaje rural en España: A la búsqueda de una racionalidad. Actas Coloq. Hispano-Francés sobre espacios rurales, 1, Madrid.

Montserrat, P. (1977). Base ecológica de las culturas rurales. Actas I Congr. Esp. Antropol. 1, Barcelona, 217–230.
Puigdefábregas, J. (1987). Transformaciones de las pautas de utilización del suelo y sus consecuencias para la gestión de los recursos renovables. J. M. Camarasa (Ed.). *El Futuro de la Gestión de los Recursos Renovables en España,* J. M. Camarasa (Ed.), programa FAST. CSIC, Madrid, 23–28.
Sltayer, R. O. (1977). Dynamic changes in terrestrial ecosystems: pattern of change, techniques for study and applications to management. MaB Technical Notes, 4. UNESCO.

14 Breeding Birds in Traditional Tree Rows and Hedges in the Central Po Valley (Province of Cremona, Northern Italy)

RICCARDO GROPPALI

Abstract: *Three rows and hedges were and, in part, still are among the most typical elements of the agricultural landscape of the central Po Valley. Nowadays, these features have lost most of their economic importance and indeed are now considered to be a nuisance or hindrance to mechanical cultivation. Tree rows as well as hedges are therefore being removed at an increasing rate. Some of the most typical tree rows and hedges still present in the central Po Valley have been studied to determine their role as breeding sites for birds and thus to evaluate their importance in agroecosystems. Finally, it is suggested that the maintenance, and possibly the reintroduction, of tree rows and hedges, even in highly productive agricultural areas, can be beneficial to bird populations.*

INTRODUCTION

Trees and shrubs are among the most threatened features of landscape physiognomy in the central Po Valley. The disappearance of tree rows and hedges along field borders is proceeding at a fast rate, as is happening in most other countries (Muir and Muir, 1987).

Trees and shrubs, planted in rows and/or hedges, were first used by the Etruscans and later by the Romans as distinct means of division between cultivated fields or pastures belonging to different owners. Dominant trees in the rows were the common maple *(Acer campestre)* and the English elm *(Ulmus minor)*, upon which the vine *(Vitis vinifera)* climbed and produced abundant grapes. Indeed this type of cultivation was called "vite maritata" (promiscuous vine growing). A dominant element in the hedges quite often was the hawthorn *(Crataegus monogyna)*, which prevented trespassing and browsing damage by cattle to cultivated fields.

This sort of landscape survived until the 18th and 19th centuries, when it proved to be more economic to introduce the cultivation of pollarded white mulberry *(Morus alba)*, whose leaves were fed to silkworm larvae (the silk industry was booming at that time). A need for quick wood production also developed and elms and maples were replaced by London planes *(Platanus hybrida)*, frequently coppiced, and black locust *(Robinia pseudacacia)* (Groppali, 1988).

At present, because of their decline in economic importance, these trees are also being eliminated, being an obstacle to field enlargements according to modern agricultural practice. For example, during 9 years only (1980–1989) on 2.430 ha of land in the municipality of Cremona, 17% of the tree rows and 30% of the hedges were eliminated. In cultivated areas, however, the percentage of eradication rises to 33 and 36%, respectively, for tree rows and hedges (Groppali, 1990) (Figure 1).

To assess how these changes may influence breeding or potentially breeding species of bird populations (Arnold, 1983; Lysaght, 1987; Murton and Westwood, 1974) studies were carried out in 1989 in two intensively cultivated parcels (0.25 km^2 each) in the central Po Valley.

In the first parcel, with 1822 m of tree rows and hedges and 13 isolated trees, 25 different bird species (1 not nesting) and 64 to 85 pairs were found. In the second parcel, with only 56 m of hedge and 7 isolated shrubs, 9 different species (6 not nesting) and 13 to 15 pairs were found (Groppali, 1989).

METHODS

To evaluate the importance of only tree rows and hedges in the central Po Valley, in 1989 and 1990 four samples of the most typical and traditional types of this element (each 250 m long) were studied in the municipality of Stagno Lombardo, province of Cremona, near Cascina Cadellora. Tree and shrub species were identified and breeding birds (pairs) were counted by using a mapping census in eight visits each year from the 15th of May to the end of June. The method consists in the localization of all birds seen or heard over the course of the visits during the breeding season (Canova, 1988), to obtain the number of nesting pairs. The results are listed below. The tree and row hedge composition is as follows:

- Mixed, dominated by common oaks *(Quercus robur)*, 20–30 m high
- Pure hybrid black poplars *(Populus x euroamericana)*, 18–25 m in height
- Pure pollarded white mulberries *(Morus alba)*, 5–6 m high
- Pure coppiced London planes *(Platanus hybrida)*, 6–10 m in height

RESULTS AND DISCUSSION

In the period under study, the above-mentioned tree rows and hedges near Stagno Lombardo (Cremona) were used for nesting by different bird pairs (maximum numbers of which are reported in Table 1).

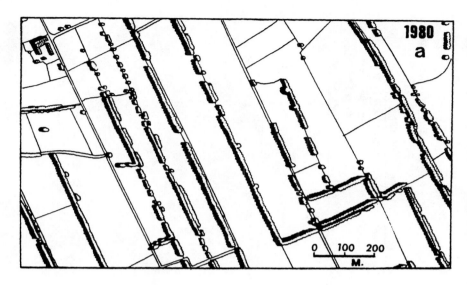

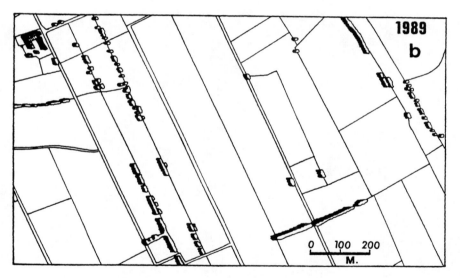

Figure 1. Example of tree removal in the central Po Valley: the same area between Gerre Borghi and Gerre de' Caprioli (S.E. of Cremona, Northern Italy) in 1980 (Figure 1a) and years later, in 1989 (Figure 1b).

Nesting bird pair computations in fixed lengths of different tree rows and hedges does not allow comparisons with apparently similar data obtained in various areas of farmland: for example, ground-breeding species are not counted.

The regular shape of the chosen shelter belts, moreover, excludes the influence of hedge intersections on the number of nesting pairs (Lack, 1988) and the local

Table 1 Birds Nesting in Traditional Tree Rows and Hedges in the Central Po Valley (Province of Cremona)

	Maximum numbers of nesting bird pairs (1989–1990, Central Po Valley) in 250 m of			
	Coppiced hedge of *Platanus hybrida*	Tree row of pollarded *Morus alba*	Tree row of *Populus x euroamericana*	Mixed tree row dominated by *Quercus robur*
Falco subbuteo	—	—	—	1
Streptopelia turtur	—	—	—	1
Upupa epops	—	1	—	—
Picus viridis	—	—	—	1
Picoides major	—	—	—	1
Luscinia megarhynchos	4	1	6	5
Turdus merula	2	1	2	2
Sylvia atricapilla	—	1	1	2
Phylloscopus collybita	—	—	—	2
Muscicapa striata	—	1	—	1
Parus major	—	—	1	3
Oriolus oriolus	1	—	—	1
Lanius collurio	—	—	—	—
Corvus corone cornix	—	1	—	3
Sturnus vulgaris	—	—	4	3
Passer montanus	—	2	12–15	17–20
Fringilla coelebs	—	1	—	1
Carduelis chloris	—	—	—	2
Carduelis carduelis	—	—	—	2
Total	7	9	26–29	49–52

scarcity of fruiting plants (except white mulberry) excludes strong interactions with frugivorous birds (Snow and Snow, 1988).

The method used, however, permits us to reliably evaluate (at least in the central Po Valley) the number of bird species and breeding pairs disappearing with the removal of measurable lengths of traditional tree rows and hedges in the cultivated landscape.

CONCLUSIONS

The results demonstrate that in intensively cultivated areas tree rows and hedges represent the only refuges for survival and reproduction for many bird species. These will soon disappear if the removal of trees and shrubs continues as in recent years.

It is also of paramount importance to improve, where possible, the "ecological value" of tree rows and hedges, in areas where these traditional features are being preserved.

Bird communities will thus benefit if a few trees are planted or are left to grow among hedge shrubs and if complete removal of shrubs in productive tree plantations along field borders is avoided.

REFERENCES

Arnold, G. W. (1983). The influence of ditch and hedgerow structure, length of hedgerows, and area of woodland and garden on bird numbers on farmland. *J. Appl. Ecol.*, 20:731–750.

Canova, L. (1988). Il mappaggio degli uccelli nidificanti in Lombardia. Dipartimento Biologia Animale dell'Università, Pavia, 11.

Groppali, R. (1988). Ambienti umidi, boschi e colture arboree negli scritti di naturalisti, geografi ed agronomi cremonesi dell'Ottocento. R. Bertoglio, V. Ferrari and R. Groppali (Eds.), Natura e ambiente nella provincia di Cremona dall'VIII al XIX secolo, Publ. Assessorato all'Ecologia della Provincia di Cremono, 99–112.

Groppali, R. (1989). Avifauna nidificante in due aree padane ad agricoltura intensiva: confronto tra un ambiente con filari e siepi ed uno privo di tale dotazione in provincia di Cremona. Proc. V. Conv. Ital. Ornitol., 4–8 ottobre 1989, Bracciano, Roma, Suppl. *Ric. Biol. Selv.*, 17:173–175.

Groppali, R. (1990). Distruzione di elementi naturalistici e paesaggistici nella Valpadana interna: l'esempio di Cremona negli anni dal 1980 al 1989. *Monti Boschi*, 41(6):14–16.

Lack, P. C. (1988). Hedge intersection and breeding bird distribution in farmland. *Bird Study*, 35:133–136.

Lysaght, L. S. (1989). Breeding bird populations of farmland in mid-west Ireland in 1987. *Bird Study*, 36:91–98.

Muir, R. and Muir, N. (1987). *Hedgerows, Their History and Wildlife.* Michael Joseph, London, 242.

Murton, R. K. and Westwood, N. J. (1974). Some effects of agricultural change on the English avifauna. *Br. Birds,* 67:41–69.

Snow, B. and Snow, D. (1988). *Birds and Berries.* T.&A. D. Poyser, Carlton, U.K. 252.

15 Bird Fauna in the Changing Agricultural Landscape

A. FARINA

Abstract: *Seasonal dynamics of birds were studied in a 100 km² rural landscape in North Italy by using GIS technology coupled to a line transect bird census. The distribution of birds is largely affected by the fine-grained mosaic of the land use. The abandonment of agricultural practices, in this region in these last decennia, has produced the decline of granivorous and fruit consumer bird species. In late autumn and in winter, birds move from the woodlands to congregate mainly in olive orchards and in fields close to hamlets and villages. During the breeding season the opposite occurs. High habitat turnover in agricultural sites is experienced by many generalist species that move from one land use to another according to the man-produced resources availability. In April rural landscapes attract many trans-Saharan birds that use small, plowed fields, hedgerows, and riparian shrublands as stopover resting and foraging sites. If in the future the abandonment trend continues, it is not unreasonable to assume that the increase of fire hazards and the spread of devastated woods would result in the consequent decline of many seasonal events as bird migrations.*

INTRODUCTION

During these last several years the abandonment of agricultural practices and transformations of land use have produced great modifications in the vegetation cover in many rural landscapes of western Europe (Farina, 1989).

To investigate the effects of these recent changes on the structure and function of the rural landscape, an ecological analysis was carried out combining non-standard bird census techniques and the Geographic Information Systems (GIS) methodology in a 10 × 10-km area of a hilly and submontane region of northern Italy. The use of birds as ecological indicators of these changes largely depends on their capacity to utilize many physical and biological cues in habitat choice and evaluation, which means that they are extremely sensitive to changes of the environmental structure (Farina, 1986a; Farina and Meschini, 1986).

This approach has allowed us to investigate the dynamic patterns of many species of birds in relation to their complicated phenology in the southern part of the west palearctic region. We evaluated bird sensitivity to the structural and functional landscape, also testing new methods of censusing birds at an unusual resolution scale for ornithological studies combined with GIS methodology (for an introduction to GIS methodology in ornithology see Shaw and Atkinson, 1990).

STUDY AREA

The study area, ranging between 200 and 900 m above sea level, is localized in the Magra watershed, on the western slope of the Apennine chain. The area is characterized by a complicated mosaic of cultivations and residual woodlots (Figure 1). The main types of cultivations are olive orchards and vineyards on the southwest slopes and alfalfa and maize fields in the river terraces of the lowlands. Chestnut orchards (a very important cultivation for human populations in the past) survive only in relict areas. In fact, a large part of these cultivations are abandoned today and have been transformed by secondary succession into dense shrublands, often subsequently destroyed by fires.

The remnant woodlots are composed of *Quercus pubescens, Q. cerris,* and *Q. robur.* This last species is restricted to the lowlands. Other common species of trees are *Ostrya carpinifolia, Fraxinus ornus,* and *Pinus pinaster* (Ferrarini, 1982). Agriculture has been reduced during these last 30 years and the abandoned fields are transformed by secondary succession into shrublands. Around the 1950s the landscape mosaic was represented by a matrix of cultivations (*sensu* Forman and Godron, 1986). Today the matrix is represented by woodlands and connectivity of wood patches is continuously in progress.

METHODS

The composition of avifauna and seasonal dynamics was investigated using the unusual spatial scale for bird censuses of 100 km^2. In fact, the standard area in which birds are usually studied does not exceed 10 to 20 ha (Ralph and Scott, 1981). Most birds have a home range wider than a single patch, and for a more realistic representation of bird distribution it is necessary to work on wider areas comprehensive of many habitats or land-use patches. The landscape fragmentation of most Mediterranean regions forces us to use areas sufficiently wide to contain most of the landscape patch types in accordance with the landscape theory (Forman and Godron, 1986; Farina 1990). The spatial scale adopted allowed us to evaluate the number of species at regional levels and the effects of complicated dynamics of bird populations typical of the Mediterranean region. The morphology and land use were introduced into a GIS (Burrough, 1986) according to the pMAP methodology (Golden Software, Springfield, WV, U.S.A.) with cells of 200 × 200 m (raster representation).

The information on birds was collected using six line transects totally 120 km, repeated three times during 1990 (February, April, and July). The transects

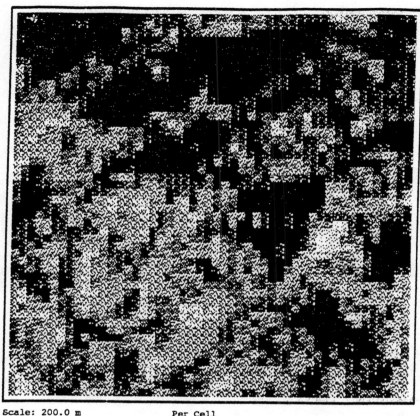

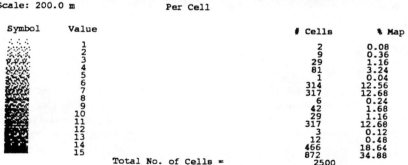

Figure 1. Distribution of land use in the study area. Gray (sparse dotted) = urban settlement; moderately dotted = agriculture; dark (densely dotted) = woodlands (see Table 1).

were made on horseback. Using this "vehicle" it was possible to work for many hours without tiring to note with attention the visual and acoustic activity of the birds and to travel at a constant speed. The location of bird contacts was noted on a map at a 1:25,000 scale and transferred to the GIS for processing.

Table 1 Definition of the Land Use Gradient According to the Importance of Each Land Use Found in a Cell of 200 × 200 m

	1	2	3	4	5	6	7	8	9	10	11	12	13	14	15
R	a	b	c	d	b	0	0	c	d	d	0	c	d	0	0
C	0	d	c	b	0	a	b	d	c	d	c	0	0	d	0
W	0	0	0	0	d	0	d	d	d	c	c	c	b	b	a

R = rural settlements, C = cultivations, W = woodlands; a = 100%, b = 75%, c = 50%, d = 25%.

A spatial and standard numerical analysis was carried out in order to ascertain the richness, the abundance, and the turnover of birds in the study area and in the component patches. The distribution and abundance of species were compared with the distribution of a habitat gradient represented by the importance of the three land use categories (rural settlements, cultivations, and woods) present in each cell (Table 1). The land use combinations were arranged along a habitat gradient from rural settlements covering the whole cell (100%, gradient class 1) to pure wood (100% of wood cover, gradient class 15).

To measure the level of heterogeneity of bird abundance in the landscape, two indexes (CH' and SH') were adopted using the Shannon diversity (Shannon and Weaver, 1949): the first index measures the diversity of clumps of equi-abundant cells CH' $= -\Sigma p_i \log p_i$, where p_i was the relative importance of the ith clump. A clump is defined as a group of cells with the same value of abundance. The second index measures the diversity of each class of clump dimension SH' $= -\Sigma k_i \log k_i$, where k_i is the relative abundance of the i_{th} class of clump abundance. The CH' index measures the fragmentation of bird abundance; the SH' measures the heterogeneity of clump dimension. The two indexes may be utilized to describe the spatial patterns of specific or population abundance (all species considered).

RESULTS

The distribution and abundance of bird species were subject to the patchy distribution of land uses. The cells with two or three land uses, and the cells dominated by cultivations, supported the higher number of individuals (Table 2) while the lowest number of individuals was found in woodlands. In Figure 2 cells are indicated with the highest value of interperiod turnover of abundance.

The habitat patches with the highest interperiod turnover are dominated by the classes of land use cover No. 6 (pure cultivations) and No. 10 (25% rural settlements, 25% cultivations, and 50% woods). The lowest value of interperiod turnover of abundance is shown by cells dominated by woods (Table 3). The CH' index showed a slight increase during April and the SH' index a decrease from winter to summer (Table 4). This behavior is quite different if we test these indexes on species and not on the all-bird populations. For example *Erithacus rubecula* had a clear decrease of both indexes from winter to summer (Table 4).

Table 2 Number of Species and Mean Number of Individuals Per Cell Along the Habitat Gradient

	Species	Individuals
1	2	3
2	14	14
3	25	29.5
4	44	26.1
6	6	29.1
7	61	25
8	23	52
9	40	36.5
10	34	26.6
11	60	19.6
12	16	38
14	60	16.7
15	56	12.6

A total of 78 species were recorded — 44 in February, 62 in April, and 58 in July (Table 5). The diversity reached the highest values in April and so did the maximum number of contacts. In February, six species were found dominant (a species is considered dominant when it has an abundance of >5% of all collection): *Passer domesticus italiae, Erithacus rubecula, Turdus merula, Parus caeruleus, Parus major,* and *Fringilla coelebs*; in April eight species: *Troglodytes troglodytes, Sylvia atricapilla, Phylloscopus collybita, Erithacus rubecula, Turdus merula, Parus major, Passer domesticus italiae,* and *Carduelis chloris*; and in July five species: *Sylvia atricapilla, Turdus merula, Passer domesticus italiae, Apus apus,* and *Hirundo rustica. Passer domesticus italiae* was found dominant in all three periods.

DISCUSSION

The study area is characterized by a fine-grained mosaic of cultivations and woods. The abandonment of agriculture has produced a change in the mosaic matrix from cultivations to woodlands. These changes are affecting the distribution and abundance of many species of birds. In fact, after this abandonment we are expecting an abrupt decline, especially in the species consuming cultivated grains and fruits (olives) and feeding in open and treed pasturelands.

During the short period of the investigation we recognized two main dynamics linked to seasons: the concentration in winter of many species of birds in olive orchards and around rural settlements and their spreading to woodlands in spring and summer. In autumn, the situation becomes similar to that in winter (Farina, 1986b, 1987).

The role of many areas of the Mediterranean as wintering places for trans-Alpine migrants and for montane circum-Mediterranean species is well known (Farina, 1987, 1988). In the study area, wintering is especially evident in olive

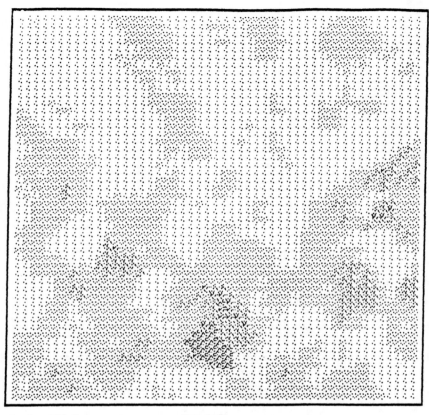

Figure 2. The distribution of cells with the higher seasonal turnover of bird abundance (all species considered). Gray, sparse dotted = low turnover; dark, densely dotted = high seasonal turnover (February–April and April–July).

orchards and along rivers (Farina, 1986b; 1988; 1991). The high interseasonal turnover, found especially in cultivated patches, probably depends on the habitat shift of many generalist species linked to agricultural activity. In April the cultivated areas become stopover sites for many species of trans-Saharan birds (e.g., *Anthus trivialis, A. spinoletta, A. pratensis, Phylloscopus collybita, P. bonelli, P. sibilatrix,* and *P. trochilus*) that select very small patches of cultivations, hedgerows, plowed fields, and riparian shrublands (Moreau, 1961, 1972;

Table 3 Distribution of the Interperiod Turnover of Abundance (ITA) Divided Into 5 Classes Along the 15 Land Use Gradients

ITA	1	2	3	4	6	7	8	9	10	11	12	14	15	Total[a]
1–10		33	44	33	34	40	33	29	48	42	45	53	72	1329
11–20	100	67	52	58	52	51	67	60	41	49	36	41	2	1021
21–30			4	4	7	7		10	6	7	18	4	1	111
31–40				4	2	1			3				1	18
>40					4									

[a] Total number of individuals per session (all transects).

Table 4 Variation of CH' and SH' Indexes of Diversity During the Periods

	CH'	Nc	SH'	Ncs
February	5.60	609	3.28	35
	(2.27)	(67)	(2.13)	(27)
April	5.97	701	3.12	30
	(1.02)	(20)	(1.01)	(12)
July	5.91	733	3.01	28
	(0.73)	(10)	(0.73)	(8)

Nc = number of clumps, Ncs = number of clump-sized patches. The values in parentheses were calculated from *Erithacus rubecula* data only.

Table 5 Number of Species (Nsp), Diversity (H') and Total Abundance (Tot) in the Selected Periods

	February	April	July
Nsp	44	62	58
H'	2.67	3.16	3.07
Tot	3299	4607	4059

Abramsky and Safriel, 1980; Farina, 1986b). The fine-grained distribution of bird abundance responds to the structure of landscape mosaic. This pattern changes slightly during the selected period despite a high interperiod abundance turnover. In other words, the species change cell position in the landscape during the year (see the example of *Erithacus rubecula*) but the general pattern of the landscape mosaic stays more or less the same when all the species are considered. In conclusion, this pattern is scale-dependent and a multiscale approach to evaluating bird dynamics is recommended, as stressed also by other authors (e.g., Wiens et al., 1986; Bennett, 1990).

The reduction of cultivated areas is producing an invasion of shrubs and trees in abandoned fields. If the reduction of agricultural practices progresses at this speed, in the next 50 years the landscape mosaic in the study area will be completely different and many of the species recorded in this study will have disappeared. Contrary to expectations, increase of woodland cover does not produce an increase of bird diversity, because in Mediterranean regions to the high frequency of fires and periodic cutting every 15 to 25 year prevents the progress of secondary succession toward mature stages.

REFERENCES

Abramsky, Z. and Safriel, U. (1980). Seasonal patterns in a Mediterranean bird community composed of transient, wintering and resident passerines. *Ornis Scand.*, 11:201–216.

Bennett, W. A. (1990). Scale of investigation and the detection of competition: an example from the house sparrow and house finch introductions in North America. *Am. Nat.*, 135:725–747.

Burrough, P. A. (1986). *Principles of Geographical Information Systems for Land Resource Assessment.* Monographs on soil and resource survey n.12. Clarendon Press, Oxford, 194.
Farina, A. (1986a). Le comunita' di uccelli come indicatori ecologici. *Atti III Conv. Orn.,* Salice Terme, 1984, 185-190.
Farina, A. (1986b). Bird communities wintering in northern Italian farmlands. *Suppl. Ric. Biol. Selvaggina,* 10:123-135.
Farina, A. (1987). Autum-winter structure of bird communities in selected habitats of central-north Italy. *Bol. Zool.,* 54:243-249.
Farina, A. (1988). Bird community structure and dynamism in spring migration in selected habitats of northern Italy. *Bol. Zool.,* 55:327-336.
Farina, A. (1989). Bird community patterns in Mediterranean farmlands: a comment. *Agric. Ecosyst. Environ.,* 27:177-181.
Farina, A. (1991). Recent changes of the mosaic pattern in a montane landscape (North Italy) and consequences on vertebrate fauna. *Option Mediterraneen* 15:121-134.
Farina, A. (1990). Rapporto tra l'ecologia del paesaggio e le altre teorie ecologiche. *Econ. Mont. Linea Ecol.,* 22(4):27-39.
Farina, A. and Meschini, E. (1986). The Tuscany breeding bird survey and the use of bird habitat description. Proc. IX Int. Conf. Bird Census and Atlas Work, Dijon, 1985. *Acta Ecol. Oecol. Gen.,* 8:145-155.
Ferrarini, E. (1982). Carta della vegetazione dell'Appennino tosco-emiliano dal Passo della Cisa al Passo delle Radici. *Bol. Mus. St. Nat.,* 2:5-26.
Forman, R. T. T. and Godron, M. (1986). *Landscape Ecology.* John Wiley & Sons, New York.
Moreau, R. E. (1961). Problems of Mediterranean-Saharan migration. *Ibis,* 103:373-427.
Moreau, R. E. (1972). *The Palearctic-African Bird Migration Systems.* Academic Press, New York.
Ralph, C. J. and Scott, J. M. (Eds.) (1981). Estimating numbers of terrestrial birds. *Studies in Avian Biology,* No. 6. Cooper Ornithological Society, Los Angeles, CA, 630.
Shannon, C. E. and Weaver, W. (1949). *The Mathematical Theory of Communication.* University of Illinois Press, Urbana, IL.
Shaw, D. M. and Atkinson, S. F. (1990). An introduction to the use of geographic information systems for ornithological research. *Condor,* 92:564-570.
Wiens, J. A., Addicott, J. F. and Case, T. J. (1986). The importance of spatial and temporal scale in ecological investigations. *Community Ecology,* J. Diamond and T. J. Case (Eds.), Harper & Row, New York, 145-153.

16 Study Outline on Ecological Methods of Afforestation

P. CORONA

Abstract: *In temperate regions, one of the major changes in land use has been brought about by the progessive abandonment of rural areas and by an increasing potential for forestry. In afforestation, it is important to generally improve the natural quality of landscape. The application of the principles of landscape ecology will allow not only the functional efficiency of afforested areas in terms of silvicultural and economic functionality, but also the ecological soundness of the resulting landscape. On the basis of research and experience carried out in temperate regions, the present study synthesizes some issues concerning ecological landscape design and forest plantation establishment through focus on spatial patterns, choice of species, site preparation, and early management operations.*

INTRODUCTION

In many temperate regions, one of the major changes in land use has been brought about by the progressive abandonment of rural areas and by an increasing potential for forestry (Bagnaresi et al., 1986). Therefore, the planning problems involved in afforestation activities and associated landscape changes are increasingly important. This is seen, for instance, in the European Community's initiatives (e.g., CCE, 1988), necessarily framed within broad-scale environmental planning (e.g., EEC Directive 337/1985).

Although landscape planning has been regarded as creating the environment by arrangement of spaces in order to withstand horizontal and vertical pressures of human impacts, the study on ecological consequences of such activity has begun only recently.

Recent developments in forestry have created more room for landscape ecological considerations (Zwolinski, 1990). The specific goals of an afforestation project may vary with the technical targets and the conditions of the project site, but it is accepted that the project should generally aim at (1) enhancing the natural qualities of the landscape by increasing diversity of the vegetation cover (such as a combination of forest stands and open spaces) and by stimulation of higher species diversity; (2) assisting in the development of a wide range of habitats; (3) eliminating visual intrusion; and (4) meeting the operational needs of plan-

Table 1 Simplified Identification of Hierarchical Levels Biologically Relevant to Forest Plantation Establishment

Hierarchical level	Spatial scale (m^2)	Temporal scale (yr)
Society	10^{10}–10^{14}	10^1–10^3
Plantation	10^4–10^6	10^1–10^2
Seedling	10^{-1}–10	10^{-1}–10
Organ	10^{-6}–10^{-2}	10^{-3}–10^{-1}
Tissue	10^{-9}–10^{-4}	10^{-6}–10^{-4}
Cell	10^{-10}–10^{-8}	10^{-9}–10^{-7}

(Modified from Margolis, H. A. and Brand, D. G. (1990), *Can. J. For. Res.*, 20(4):375–390. With permission.)

tation management, taking into account minimalization of planting operation impact on soil.

In practice, in many situations (4) overrides the other factors. It is important to observe that these options are space- and time-oriented and also that forest plantations, like all biological systems, can be regarded as hierarchical in organization (Margolis and Brand, 1990), with each hierarchical level operating on a particular spatial and temporal scale (Table 1).

From an operative point of view, the main factors of forest plantation establishment directly related to ecological landscape design can be outlined as (1) *spatial pattern* (e.g., plantation geometry and the retention of open spaces); (2) *stand effects* (e.g., choice of species and species grouping); and (3) *system effects* (e.g., the techniques of soil preparation and the maintenace of fertility).

On the basis of research and experience carried out in temperate regions, the present study aims at synthesizing some issues concerning these topics by focusing on afforestation spatial pattern, choice of species, site preparation, and early management operations.

AFFORESTATION SPATIAL PATTERN

Managing forest plantations as a complex, multipurpose component of the countryside involves landscape ecological concepts which refer to structure and pattern (Anko, 1990).

Biodiversity is also an important feature and, indeed, afforestation is a useful means to introduce diversity into rural and suburban landscapes. Within forest plantations, diversity can be achieved by the use of various species groups (see next section), by establishing adjacent areas at different times, so as to have uneven-aged groups, and by making appropriate use of unplanted open spaces (these are of great value for ecological and visual diversity).

The ecological value of plantation edges (ecotones) is generally greater than the values of the interior. In manmade woodlands, visual links with adjacent areas should be made as natural as practicable and can be improved by structural differentiation of the edges, with strong indentations and protrusions. Species with different growth rates can be used to vary tree height and can be established in irregular outlying groups, blending a mixture of hardwood overstory and understory and small shrubs into adjacent meadows. The natural bankside vegetation near watercourses with unplanted irregular streamside strips can also be maintained (Forestry Commission, 1988, 1989).

The landscape value of a forest plantation on a spatial scale is also related to its shape, with the distinction between natural and geometric shapes being especially significant and playing a major role in landscape design. Right angles, symmetrical shapes, and long, straight edges should be avoided. If the plantation geometry strictly follows the cadastral boundaries, it may cause an extraneous effect in respect to the surrounding scenery: plantation edges have to be maintained away from property boundaries. In hilly and mountain environments, the upper margin of afforested areas is usually the most prominent and it is often expedient not to plant on it, thus stressing visual dominancy (Corona, 1989).

CHOICE OF SPECIES

The ecological efficiency of afforestation is closely related to dendrological composition. Obviously, the tree species to be planted must suit the microclimatic and soil conditions of the site to satisfy the technical targets of the afforestation project. The ability of selected tree species to buffer changes in environmental conditions is determined by the ecological amplitude of the species and its ability to survive and grow under certain ranges of environmental conditions.

The species pattern of forest plantations should reflect the broader pattern of the landscape, with a specific reference to natural woodlands. Taking into account direct and indirect effects which can have ecological and economic significance, it seems best to follow naturally dominant features and, therefore, it is very important to fully understand the natural communities that could develop in any proposed project site.

The concept of variability can be widely used and can be produced by planting mixtures of groups of different species and age. In temperate regions, a high level of diversity is phytosociologically possible even if few tree species are dominant (Austin, 1984). Groups, rather than individuals, of one species can be planted into the mass of another: an overlapping and interlocking pattern of species groups, distributed in relation to site conditions, gives overall natural unity. To enhance diversity, some trees and shrubs with showy flowers and/or edible fruits to support wildlife may be placed towards the ecotones (Corona, 1989).

In the long term, some species, especially fast-growing ones (e.g., some eucalypts, poplars, and pines), may modify the physical and chemical soil properties (Whitehead, 1982). Such species may assimilate nutrients at a high rate and may contribute to soil impoverishment (Ciancio et al., 1983).

Fast-growing conifer plantations have shown longer times to reestablish an equilibrium than natural woods (Swank, 1982; Zwolinski, 1990). In New Zealand, productivity decline occurred because of nutrient impoverishment in plantations cultivated for more than one rotation with one tree species (*Pinus radiata* D.Don). The soil quality in those plantations would not allow sustainable biomass production at a high rate. This kind of productivity decline could be caused by autoinhibition phenomena: toxins exuded by roots or trees of the previous rotation (allelopathic phenomena) may have an effect on decomposition cycles and organic matter mineralization. Such phenomena, however, rarely seem to be determinant for inhabiting edaphic processes in forest plantations of temperate regions, which, at the end of the rotation, can almost always support a significant humic layer (Lucci, 1987).

SITE PREPARATION

The mechanized operations for site preparation before tree planting have the objective of improving soil conditions for successful forest establishment, but they have to be carefully handled in order to prevent negative environmental effects (Tew et al., 1986).

Among the operations before soil preparation, land clearance makes subsequent tillage easier and reduces the competition among plants for water, nutrients, and sometimes, light, but, since it is often done by a bulldozer, tends to involve a partial removal of residual vegetation and of organic soil horizons (Lucci, 1987). Where risks of slope instability or erosion are high, land clearance can be done on localized strips. Cover should be left in vulnerable areas, e.g., near watercourses to reduce the effects of sedimentation (Forestry Commission, 1988). When the whole surface is cleared, brushpiles can be aligned on the contour, aiming to limit both rill and gully erosion. If a bulldozer rake is used rather than the blade, the removal of organic matter can be reduced (Lucci, 1987). Land clearance by chain or hammer strippers can maintain *in situ* more organic matter, which subsequently will be imbided by the soil in tillage operations. In practice, there are difficulties in the availability of these machines, caused by frequent adverse site conditions (e.g., stoniness, etc.) or by limits in the tractors at disposal (Eccher et al., 1983); moreover, the roots remaining in the soil tend to produce stong vegetation regrowth, thus hampering tillage.

As regards soil preparation for forest planting, deep tillage, like furrow plowing, is probably the most commonly used technique, at least in temperate regions. Many authors have stated its positive effects in establishing forest plantations, but there is less agreement about long-term results on plantation productivity.

Actually, deep tillage is often a landscape change perceived to be environmentally intrusive (Corona and Leone, 1988). Deep plowing has potential risks with respect to the environment, namely: disturbance of slope stability; stimulation of diffuse erosion processes; organic matter removal or reduction; organic matter displacement into deeper and less active horizons and as well as its distribution into a bigger soil volume, dispersing its nutritive potential (which is very dangerous, especially if carried out in previously untilled land); alteration of soil water balance; alteration of soil biological activity; accentuation of mineralization and alteration of nutrient cycles; and increased leaching (Lucci, 1987). Water quality may also be affected by increased turbidity and eutrophication, thus potentially damaging ichtyofauna. Increased sedimentation may also alter the hydromechanical dynamics of the watercourses.

Control of the negative effects of deep plowing can be accomplished (Loreti and Pisani, 1986; Forestry Commission, 1988) as follows:

1. Plow only where it is necessary, choosing alternative tillage techniques (e.g., chiseling) to limit the displacement of organic matter or reduce the mixing up of upper soil horizons, eventually improving soil accumulation along the planting rows (e.g., by bedding) where tree roots are going to develop to a greater extent;
2. When plowing, provide cutoff drains so that individual furrows are shortened, aligning drains with gradients of not more than 2 to 5% in slope and stopping plow furrows and drain ends well short of watercourses;
3. Furrows follow natural changes of the slope in gentle curves, not at varying angles to the slope, and avoid straight alignments even on short cutoff drains; and
4. Hydrologic measures (e.g., settling silt traps to trap debris and sediments) are taken on slopes, together with tillage along the contour lines, when operationally possible.

There are many types of agricultural chisels to improve soil preparation in afforestation. The adoption of vibrating rippers, application of lateral appendices (little mouldboards) on ripper tines, adoption of subsoilers, etc., can achieve good tillage and support water infiltration and accumulation (Manfredi and Baraldi, 1986), especially in areas with a dry season like the Mediterranean, guaranteeing a good substratum for root growth, without the drawbacks of deep plowing.

Soil preparation on slopes up to 18 to 22° can be preferably done by contour tillage (on the whole surface or localized); on slopes up to 27 to 31°, tillage is practicable by slightly modifying the slope profile, making contour ridges to reduce local inclination. On very inclined slopes or in unfavorable morphologic conditions (e.g., large rocky outcrops), tillage is possible by making little terraces by shovel excavators or, where possible, by small caterpillars (Lucci, 1987).

Whatever the method, it is necessary to avoid large terraces, which often happen in practice; for instance, such as many afforestation projects in Mediterranean areas where the excessive digging causes rocks to be exposed with loss of fertile soil horizons.

EARLY MANAGEMENT OPERATIONS

The selection among the various management systems has to be related to the site conditions and to the primary tillage systems. In temperate regions, the first technical problem is weed control, since competitive vegetation can generally suppress the trees for a minimum of 2 to 4 years after planting.

Any use of herbicides has proved ecologically unsustainable in tending forest stands. When there are no obstructions, weed control can effectively be carried out by secondary tillage: for instance, by disc harrowing. Where there are stumps, rocky outcrops, or large stones, weeds can be controlled by mowing and chopping competitive vegetation. The intensity of the intervention (on the whole surface or localized) has to be defined in relation to the erosion risks. Generally, it is not recommended to carry out secondary tillage along the maximum slope lines: it is better to harrow along the contour among planting rows, at least up to slopes which can be traveled over by machinery, and to complete the intervention by localized tillage around each tree (Eccher et al., 1983).

Practical interest is growing in management options which reduce mechanical weeding operations by adopting alternative methods (Frochot et al., 1990; Zwolinski, 1990), e.g., localized piro-weeding and mulching. The latter seems to offer the best alternatives to tree growers looking for ecologically sound ways to suppress weeds and protect the long-term health of the soil, while maintaining wood yields. At present, removable plastic films are the most frequently used mulches in plantation silviculture, but organic dead mulches (e.g., wood chips, agricultural residuals, and urban wastes) are gaining in importance. Promising techniques are based on living mulches forming a weed-suppressing carpet into which trees are planted, since they can directly favor the biological efficiency of afforestation: for instance, the use of N-fixating species (e.g., subterranean clover, a self-seeding winter annual legume) not only can hamper weeds, but also gives a benefit to soil fertility without significant water competition effects on planted tree seedlings even on dry sites, since the subterranean clover dries up during the summer. Also, the presence of a cover crop reduces erosive potential, which is high immediately after tree planting (at least, in the climatic conditions of many temperate regions), and makes the retention of nutrients possible.

CONCLUDING REMARKS

Landscape is a concept as well as a biological phenomenon where forestry and ecology interact. The fate and role of afforestation in a given landscape in many ways reflect its divergence from the primeval natural landscape. A natural forest landscape comprises ecologically placed natural forest species and does not have formal human manipulation. Any tentative duplication of these processes by afforestation does not lead to a natural landscape, but, at least, it should aim at a natural arrangement (Figure 1). Afforestation efficiency should refer not only to silvicultural functionality, but also to environmental patterns (e.g., patches, corridors, and mixtures). It should take into account that the different ecosystem units tend to form an ecological mosaic both in space and time and also the dynamic behavior of the ecosystems (e.g., resource partitioning; Viola, 1981).

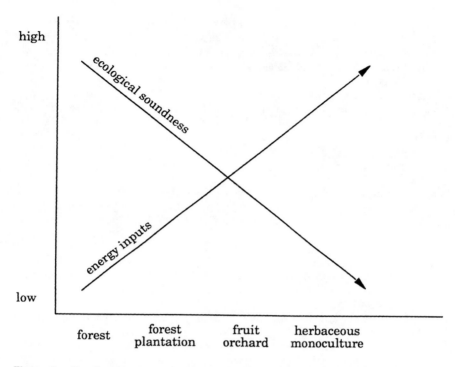

Figure 1. Functional assessment of some land uses within agroecosystems.

Comprehensive plans are therefore necessary when new planting is undertaken on a substantial scale. The operation strategies for forest plantation development need integrated planning with the components and functions of the landscape: for instance, ecological landscape designing within afforestation planning widely overlaps with soil conservation planning. From a practical point of view, both types of components (biological and technical) have to be taken into account as regards the planning process. Technical decisions always require an evaluation of the biological options and some quantitative estimate of what different afforesting regimes will ecologically and economically cost vs. what will be obtained, considering that the landscape pattern established by afforestation may persist for a long time.

REFERENCES

Anko, B. (1990). Landscape ecology in forestry: a new challenge. *Proc. 19th IUFRO World Congr.*, Montreal, Div. 1(1):149–156.

Austin, R. L. (1984). *Designing the Natural Landscape*. Van Nostrand Reinhold, New York, 117.

Bagnaresi, U., Ciancio, O., Eccher, A., Minotta, G., Pettenella, D. and Ponticelli, P. (1986). Il miglioramento dei boschi e il recupero alla produzione legnosa dei terreni agricoli abbandonati nella collina italiana. *Studio Generale della Collina Italiana*, Ass. Naz. Bonifiche (Ed.), Edagricole, Bologna, 61–84.

Commission des Communautes Européennes (1988). Stratégie et action de la Communauté dans le secteur forestier. *Com* (88) 255, Bruxelles, 57.

Ciancio, O., Mercurio, R. and Nocentini, S. (1983). Le specie forestali esotiche nella selvicoltura italiana. Annali Istituto Sperimentale per la Selvicoltura, 12/13:1–714.

Corona, P. (1989). L'Assetto Paesaggistico dei Rimboschimenti. Note Techniche SAF, 6:1–9.

Corona, P. and Leone, M. (1988). Impatti ambientali di tecniche agricole e forestali antiecologiche. *Linea Ecol.*, 20(5):63–66.

Forestry Commission (1988). Forests and Water Guidelines. Forestry Commission, Edinburgh, 40.

Forestry Commission (1989). Forest Landscape Design Guidelines. Forestry Commission, Edinburgh, 36.

Eccher, A., Scarpa, M. and Pizzedaz, S. (1983). Le tecniche di rimboschimento. *Ital. Agric.*, 120(4):128–139.

Frochot, H., Dohrenbusch, A. and Reineke, H. (1990). Forest weed management: recent developments in France and Germany. Proc. 19th IUFRO World Congr. Montreal, Div. 1(2):290–299.

Loreti, F. and Pisani, P. L. (1986). Lavorazioni del terreno negli arboreti. *Riv. Agron.*, 20(2–3):134–152.

Lucci, S. (1987). Meccanizzazione della attività forestali in relazione alle esigenze di conservazione del suolo. *Conservazione del Suolo e Forestazione in Calabria*, Laruffa Publisher, Reggio Calabria, 150–165.

Manfredi, E. and Baraldi, G. (1986). Aspetti meccanici ed energetici della lavorazione del terreno. *Riv. Agron.* 20(2–3):153–165.

Margolis, H. A. and Brand, D. G. (1990). An ecophysiological basis for understanding plantation establishment. *Can. J. For. Res.*, 20(4):375–390.

Swank, W. T. (1982). Effetti della preparazione del suolo al rimboschimento. *Econ. Mont.*, 14(6):21–24.

Tew, D. T., Morris, L. A., Allen, K. L. and Wells, C. G. (1986). Estimates of nutrient removal, displacement and loss resulting from harvest and site preparation of a *Pinus taeda* plantation in the Piedmont of North Carolina. *For. Ecol. Manage.*, 15(4):257–267.

Whitehead, D. (1982). Ecological aspects of natural and plantation forests. *For. Abstr.*, 43(10):615–624.

Viola, F. (1981). Ecologia e pianificazione della colture legnose. *Ital. Agric.*, 117(1):73–92.

Zwolinski, J. B. (1990). Intensive silviculture and yield stability in tree plantations: an ecological perspective. *S. Afr. For. J.*, 155:33–36.

17 Research on Germplasm of Herbaceous Plants in Tuscany

F. CASTIONI
G. CERRETELLI
A. DE MEO
D. NOTA
S. PADERI
M. RIGHINI
G. TARTONI
C. VAZZANA

Abstract: *The conservation of germplasm resources is important in countries such as Italy where variation in pedoclimatic and sociocultural aspects make it possible to select local populations of cultivated species especially adapted to environmental characteristics. Unfortunately, with modern agriculture, most of a crop's genetic variability has been lost. Recently, to face the problem of genetic erosion, regional areas of limited extension, where traditional agriculture is still practiced by elderly farmers, have been selected in Tuscany. Different harvesting routes have been explored by researchers to collect seeds and other germplasm material. Although the loss of genetic information has been very important in the last few decades, it is still possible to find local varieties which hold important agricultural potential. A hypothesis of active conservation of the harvest germplasm and of the constitution of a regional bank is discussed.*

INTRODUCTION

The diversification of the varieties of cultivated species has represented an important event for the diffusion of agriculture and of cultivated plants in different environments and latitudes. Man, using local and imported vegetable matter, has brought about genetic selection favoring those subjects which are better suited to the agronomical and pedoclimatic conditions.

With the advent of industrialized agriculture and the availability of fertilizers, pesticides, and hybrid seeds, traditional patterns of improvement and selection of local seeds have collapsed (Hansen et al., 1986; Sasson, 1988). Hybrid seeds have established themselves on the market mostly because of their elevated productivity, which is often directly connected to a high energy expenditure for fertilizers, pesticides, and the use of mechanization (Shulman, 1986; Kloppenburg and Kleinman, 1987). With the passing of the years, this management of agricultural methods has highlighted the problems associated with the depletion of soil resources as well as an agriculture ever more dependent on elevated energy expenditure. There has been an impoverishment of genetic variability and surplus agricultural production. In such a situation, the recovery of the germplasm of cultivated species has become important in order to:

- Restore genetic information matter,
- Restore populations
- Use to full advantage the local varieties,
- Implement genetic variability which represents a fundamental condition for the safeguarding of the agroecosystem in that intraspecific and varietal diversity is a guarantee of equilibrium, and
- Create the possibility of maintaining genetic variability which could represent an important source for the production of new varieties suited to different environments of cultivation (Plucknett et al., 1987).

METHODOLOGY

The region of Tuscany, Italy is made up of territories with very different historical, social, and environmental evolutions. It is an area with economical and technical characteristics which differ in many points of view, other than just the agricultural aspect. For this reason it was felt that it would be appropriate to evaluate the possibility of the ancient cultivars of herbaceous plants grown locally, both in those areas where there has been a lesser expansion of industrialized agriculture, and in those areas characterized by a more intense agricultural activity. The areas taken into consideration were the Garfagnana, the Amiata, and the Maremma Grossetana.

The recovery of the germplasm was brought about through widespread research across the territory. Information was obtained from agricultural offices, from public corporations, and above all from local farmers, in order to recreate a map of agricultural practices in the last decade.

In the areas of research, only vegetable matter reproduced *in loco* for at least 20 years was gathered, in order to obtain a greater guarantee of the quality of local varieties. As far as was possible, we tried to collect samplings representative of the genetic variability of the populations. At the moment of collection, information was also gathered by interviewing the farmer regarding farming methods, the ecological needs of the species, and the uses (also alimentary) connected to it. On many occasions, we became aware of the extent to which the cultivation of local plants is tied to a precise local culture on its way to becoming extinct or, uniformly at times, recovered only in its more aesthetic forms.

The samples thus obtained were collected and catalogued with all the fundamental information, and then preserved at a low temperature (5°C).

Table 1 Number of Germplasm Samples Collected Per Species in the Study Area

Species	No. of samples	Species	No. of samples	Species	No. of samples
Basil	1	Chard	2	Cucumber	4
Onion	1	Melon	2	Clover	4
Watermelon	1	Pea	2	Squash	6
Alfalfa	1	Parsley	2	Chickpea	9
Field bean	1	Sweet vetch	2	Corn	10
Lentil	1	Faba bean	2	Lettuce	10
Radish	1	Potato	2	String bean	12
Sorghum	1	Rye	2	Tomato	23
Turnip	1	Celery	3	Bean	36
Rye-grass	1	T. spelta	3		

RESULTS

The research has given satisfactory results on the presence of local landraces, with about 150 samples of vegetables collected. The vegetables found were, for the most part, horticultural (Table 1). Of these, the majority were found in the territory of Maremma, or in those more internal regions characterized by a less pronounced agricultural and economic development (Figure 1). This highlights the fact that the safeguarding of such material was preserved by small and elderly cultivators, tied to a traditional rural tradition in which the practice of the reproduction of seeds was fundamental to agricultural activity.

DISCUSSION

The material gathered holds interesting potential and offers the possibility of utilization, from agricultural, economic, and social considerations.

First, it is necessary to reproduce the material (some samples have already been replicated) and to study it in order to be able to define structural and physiological characteristics more precisely. It will then be possible to supply a description and a classification of those varieties or ecotypes gathered, and to understand if some of these were cultivated on a large scale (perhaps with different names in different areas) or whether they are local ecotypes restricted to a small area.

It is indispensable to create a germplasm bank to guarantee the preservation of the samples collected (during the program for 1991), but at the same time to maintain the incentive for active preservation on the part of interested parties, which would constitute a sort of local living bank.

The local varieties collected have to continue to circulate among the farmers. Above all, those better identified and studied samples which possess the characteristics of major resistance to adverse conditions could then contribute to a process of transformation of agriculture aimed at lowering energy and pesticide consumption because of their adaptation to lower nutrient levels.

Figure 1. Sampling areas chosen in Tuscany: ● Garfagnana; ▲ Amiata; and ⣿ Marema.

CONCLUSIONS

Although the loss of germplasm has been widespread in the last few decades, some local varieties can still be found. It must be specified, however, that such possibilities are tied to the activities of elderly gardeners or farmers and are not found among the younger people. Therefore, it is urgent to recover and safeguard vegetable populations which have now become rare in the region under study.

ACKNOWLEDGMENTS

We would like to thank the Regione Toscana for financial support, the Botanical Garden in Lucca and the Centro Sperimentale per il Florovivaismo in

Capannori for the help offered during the research. We are especially grateful to all those farmers who gave us assistance, information, and the germplasm they were able to preserve in the areas under study.

REFERENCES

Kloppenburg, J., Jr. and Kleinman, D. L. (1987). The plant germoplasm controversy. *Bioscience,* (3)3:190–198.
Hansen, M., Busch, L., Burknardt, J., Lacy, W. B. and Lacy, L. R. (1986). Plant breeding and biotechnology. *Bioscience,* (3)1:29–39.
Plucknett, D. L., Smith, N. J. H., Williams, J. T. and Murthi Anishetti, N. (1987). *Gene Banks and the World's Food.* Princeton University Press, Princeton, NJ.
Shulman, S. (1986). Seeds of controversy. *Bioscience,* (3)10:647–651.
Sasson, A. (1988). *Biotechnologies and Development.* UNESCO, Paris.

18 Grapevine Germplasm Diversity and Conservation

A. SCIENZA
O. FAILLA
L. VALENTI

Abstract: *Viticulture in Italy is one of the most ancient (over 4000 years old) and important (more than 1,000,000 ha) cultivations. In spite of this, little research has been conducted on germplasm collection, description, and conservation. With this purpose, in 1983 a research program was initiated, in collaboration with other institutes, having as its main objectives:*

1. *Research and collection of varieties cultivated in the past which are now uncommon;*
2. *Research and collection of wild grapevine (Vitis vinifera ssp. silvestris);*
3. *Collection of varieties cultivated in other regions and countries;*
4. *Description by ampelographic, ampelometric, and chemotaxonomic (seed, pollen, and root tip protein electrophoresis and anthocyanin berry skin composition) methods.*
5. *Use of numerical taxonomy methods to study the phenetic and phylogenetic relations among grape varieties, subspecies, and species.*

Such work will allow the preservation of important genetic resources. Preliminary results have shown the possibility of distinguishing objectively between both species and varietal groups.

INTRODUCTION

Viticulture in Italy is one of the most ancient types of cultivation in the world. The first evidence of grapevine cultivation dates back to 2000 B.C. in the southern part of the country and to 1000 B.C. in the northern part (Sereni, 1986). Nowadays grapes are grown on more than 1,000,000 ha of land, mainly for wine production. Recently a special grant of the Italian Council for Research (Consiglio Nazionale della Ricerche, 1988) listed more than 2200 different varieties of *Vitis vinifera* and other grape species more or less intensively grown in our country as wine, table or rootstock cultivars, or simply collected in germplasm and variety collections. Negrul (Levadoux, 1956) classified the grapevine varieties in three *proles* in relation to their phenotypical characteristics and origin zones: *orientalis, occidentalis,* and *pontica.*

According to Rives (1971) and to Sereni (1986) the French and Italian native varieties are probably the result of the domestication of the indigenous wild grapevines (*Vitis vinifera* ssp. *silvestris*), which seem to consist of the selection of the rare hermaphrodite wild vines and of spontaneous crosses between the wild vines and the *orientalis* and *pontica* varieties introduced into Italy by the Greek colonists. So, from the phylogenetic point of view, the Italian grapevine varieties could be classified in three main groups: (1) closely linked to wild vines (cvs Lambruschi, Refoschi, Teroldego, Nebbiolo, etc.); (2) closely linked to eastern varieties (cvs Moscati, Malvasie, Schiave, table cultivars ....); and (3) an intermediate group.

To study the phylogenesis of grapevine varieties and to preserve important germplasm, a research program was started in 1983. Its main objectives are

1. To research and collect varieties cultivated in the past which are now uncommon;
2. To research and collect wild grapevine (*Vitis vinifera* ssp. *silvestris*);
3. To collect varieties cultivated in other regions and countries;
4. To describe the grapevine genotypes by ampelographic, ampelometric, and chemotaxonomic (seed, pollen, and root tip protein electrophoresis and anthocyanin berry skin composition) methods, and
5. To use numerical taxonomy methods to study the phenotypic and phylogenetic relationships among grape varieties, subspecies, and species.

MATERIALS AND METHODS

Vine collections were carried out by local surveys and by asking for materials from other Italian, European, and extra-European institutes and organizations. Collections were established growing at least three plants per genotype and, when possible, the same genotype was planted in different locations as regards soil and climate. Different clones of the same variety were collected, particularly for the older and less uniform varieties.

Studies on wild grapevine distribution and characteristics were carried out as described by Scienza et al. (1986, 1989) and Anzani et al. (1990).

Ampelographic studies were carried out using the Office International de la Vigne et du Vin (1985) descriptors. Ampelometric (leaf and seed morphology) studies were carried out by the methods described by Anzani et al. (1990).

Chemotaxonomic studies were carried out by seed and pollen protein electrophoresis as described in Scienza et al. (1990b) and by anthocyanin composition of the berry skin (in colored varieties) as described by Mattivi et al. (1990).

Multiple correlation, and discriminant and cluster analysis were the main multivariate statistical methods used for numerical taxonomy.

Table 1 Organizations and Institutes Involved in the Research Program, Surface Area Cultivated For Germplasm Collection, and Number of Varieties Collected

Institutes and Organization	Surface (ha)	Varieties
Istituto Coltivazioni Arboree Milano (c/o CI.VI.FRU.CE. Voghera (PV)	2.0	1,000
Istituto Agrario Provinciale S. Michele a/A (Trento)	2.5	1,900
Centro Vitivinicolo Provinciale Brescia	0.5	200
Private farms in Pavia, Bergamo, Mantova, Siena, and other districts	7.5	2,000

Table 2 Native Origin Zones of the Grapevine Varieties and Clones Collected For the Research Program

Countries	Native origin zones
Vitis vinifera sativa	
Italy	Oltrepò pavese, Franciacorta, Riviera del Garda bresciana, Bergamo, Friuli, Alto mantovano, Basso mantovano, Trentino, Piemonte
France	Bourgogne, Champagne, Bordolais, Midì, Val de Loire
Germany, Switzerland, Spain, U.S.S.R., Hungary, Yugoslavia, Greece, Austria, and Portugal	
Vitis vinifera silvestris	
Italy, France, Switzerland, Germany, Hungary, and Northern Africa	
Vitis spp. (species, rootstocks, hybrids)	
European, Asian, and American	

RESULTS

A total of 10 grapevine collections from 5 locations in northern Italy and 1 in central Italy have been established with about 5000 different genotypes of *V. vinifera* (4500), *V. vinifera* ssp. *silvestris* (150), and other American species (500), for a total of about 20,000 vines (Tables 1 and 2).

Discriminant analyses were generally able to distinguish different varieties on the basis of their leaf or seed morphology (Figure 1); discriminant analyses were also able to distinguish the groups of genotypes made on the same basis by the cluster analysis (Figure 2); cluster analysis was able to group different genotypes

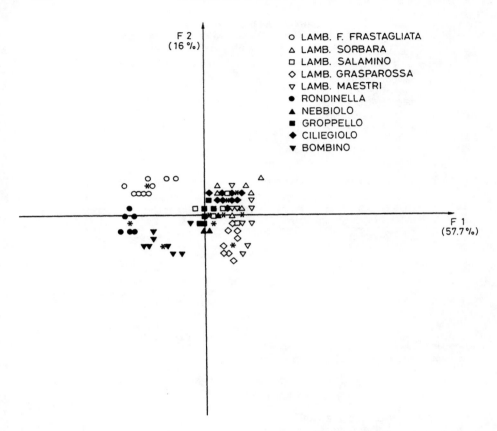

Figure 1. Distribution of 100 leaves of 10 different grapevine varieties (see the list in the diagram) in relationship in the first 2 discriminant functions (F1 and F2) calculated with 13 indexes of leaf shape, lobature, and size. The two functions explained ca. 74% of the variability, and almost 100% of the leaves were attributed to the correct variety (From Scienza, A., Failla, O., Anzani, R., Mattivi, F., Villa, P. L., Gianazza, E., Tedesco, G. and Benetti, U. [1990a]. *Vignevini*, 17(9):25–36. With permission.)

on the basis of their similarity in leaf or seed morphology (Figures 3 and 4) (Anzani et al., 1990; Scienza et al., 1990a).

Grapevine varieties showed different anthocyanin profiles of the berry skin (Figure 5) and cluster analysis was able to group different varieties on the basis of their similarity (Figure 6); and discriminant analysis could distinguish the groups of genotypes made by the cluster analysis (Figure 7) (Mattivi et al., 1990; Scienza et al., 1990a).

Grapevine varieties showed different electrophoretic patterns of seed protein (Figure 8) and cluster analysis was able to group different varieties on the basis of their similarity (Figure 9) (Scienza et al., 1990a).

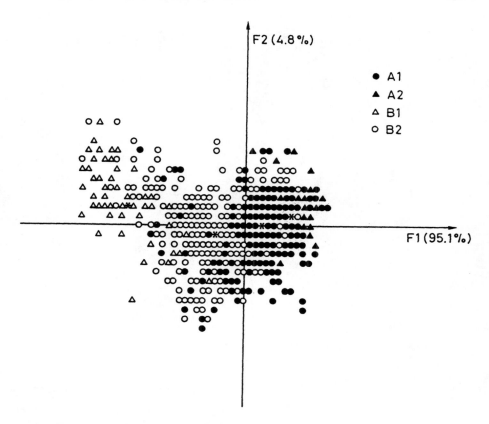

Figure 2. Distribution on the first two discriminant functions (F1 and F2) of 1370 leaves of 137 different wild vines. The vines were grouped by cluster analysis in four groups in relation to their lobature. The functions were calculated with three indexes of leaf lobature. The two functions explained ca. 100% of the variability, and almost 62% of the leaves were attributed to the correct group. A1: 5–7 lobes, A2: 5–7 deep lobe, B1: 3 light lobes, and B2: 3 lobes. (From Anzani, R., Failla, O., Scienza, A. and Campostrini, F. [1990]. *Vitis,* Special Issue, 97–112. With permission.)

DISCUSSION AND CONCLUSION

The set of variables used to describe the grapevine genotypes, and the numerical taxonomy methods adopted, proved useful to distinguish different genotypes and to group them in relation to their similarity. Frequently, these classifications did not agree with the origin and the more general aspect of the genotypes. However, we have not yet tried to use the whole set of variables at the same time nor all the variables that we think are important to classify the grape genotypes.

The preliminary results of the research program are promising and important germplasm preservation has been started.

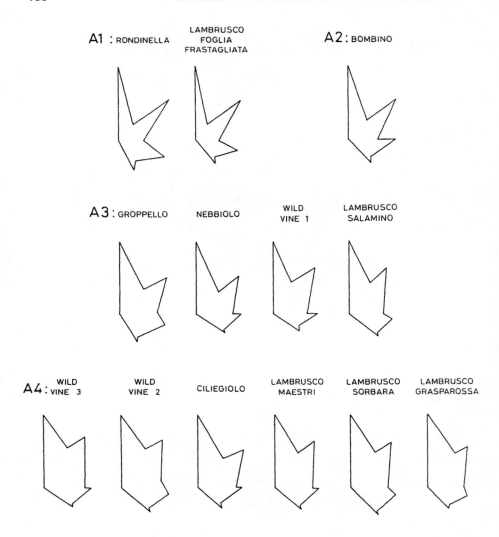

Figure 3. Half-leaf profiles of ten grapevine varieties and three wild grapevine plants as grouped by cluster analysis. Cluster analysis was carried out with three indexes of leaf lobature. A1: seven deep lobes, A2: five deep lobes, A3: five lobes, and A4: three lobes (From Scienza, A., Failla, O., Anzani, R., Mattivi, F., Villa, P. L., Gianazza, E., Tedesco, G. and Benetti, U., [1990a]. *Vignevini*, 17(9):25–36. With permission.)

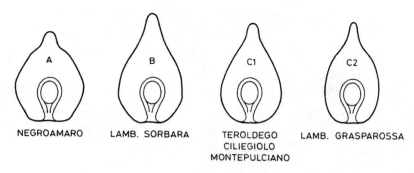
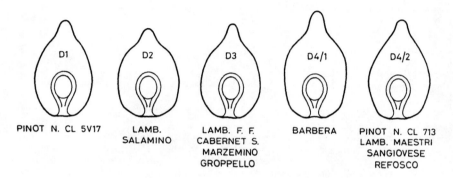

Figure 4. Seed profiles of 18 grapevine varieties as grouped by cluster analysis. Cluster analysis was carried out with three indexes of seed size and shape. A: very short, large, and round; B: very long, very enlongated; C1: narrow and enlongated; C2: very long, very narrow, and very enlongated; D1: large and round; D2: very short, narrow, and round; D3: short; D4: long and large. (From Scienza, A., Failla, O., Anzani, R., Mattivi, F., Villa, P. L., Gianazza, E., Tedesco, G. and Benetti, U., [1990a]. *Vignevini,* 17(9):25–36. With permission.)

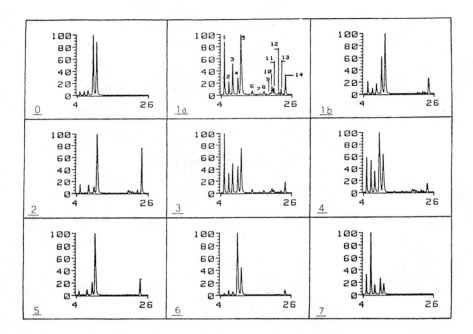

Figure 5. Nine main chromatographic patterns of anthocyanin pigments extracted from grape berry skin in the x axes: time = minutes; 1 = delphinidin-3-monoglucoside; 2 = cyanidin-3-monoglucoside; 3 = petunidin-3-monoglucoside; 4 = peonidin-3-monoglucoside; 5 = malvidin-3-monoglucoside; (6–10) = (1–5)-acetates; (11–14) = (1–5)-*p*-coumarates. (From Mattivi, F., Scienza, A., Failla, O., Villa, P. L., Anzani, R., Tedesco, G., Gianazza, E. and Righetti, P. G. [1990]. *Vitis,* Special Issue, 119–133. With permission.)

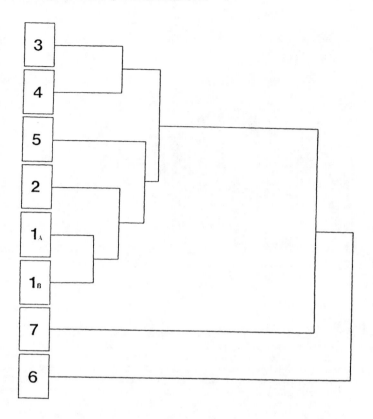

Figure 6. Dendrogram showing the results of cluster analysis carried out with seven variables relative to the chromatographic patterns of anthocyanin pigments extracted from grape berry skin. Cluster analysis was carried out on ca. 120. varieties; each variety could be classified according to the main nine chromatographic patterns reported in the Figure 5 (varieties of the group 0 were excluded by the analysis) (from Mattivi, F., Scienza, A., Failla, O., Villa, P. L., Anzani, R., Tedesco, G., Gianazza, E. and Righetti, P. G. [1990]. *Vitis,* Special Issue, 119–133. With permission.)

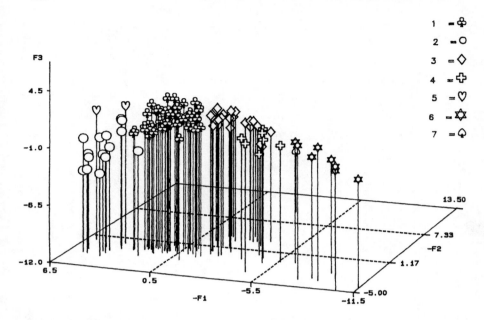

Figure 7. Distribution on the first three discriminant functions (F1, F2, and F3) of 120 grapevine varieties. The varieties were grouped by cluster analysis in 7 groups according to their chromatographic patterns of anthocyanin pigments extracted from berry skin (the 7 patterns are reported in Figure 5 — group 0 was excluded and the groups 1a and 1b were unified). The functions were calculated with 7 variables. The three functions explained 98% ca of the variability, and almost 90% of the varieties was attributed to the correct group (from Mattivi, F., Scienza, A., Failla, O., Villa, P. L., Anzani, R., Tedesco, G., Gianazza, E. and Righetti, P. G. [1990]. *Vitis,* Special Issue, 119–133. With permission.)

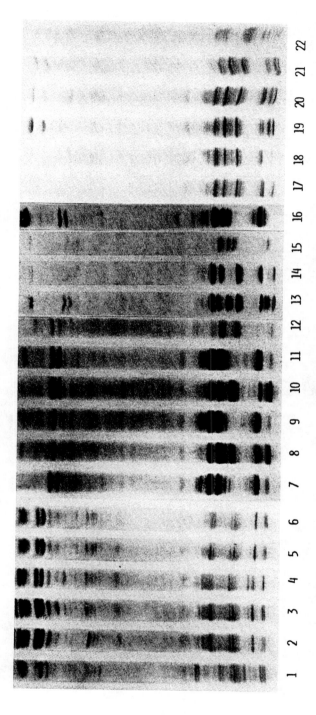

Figure 8. Electrophoretic patterns of 16 grapevine varieties and 6 wild vines of the seed endosperm proteic subunits. 1: Rosetta, 2: Rondinella 77, 3: Rondinella, 4: Dindarella, 5: Corvina 7, 7: Lambrusco di Sorbara, 8: Lambrusco Salamino, 9: Lambrusco Oliva, 10: Lambrusco Marani, 11: Lambrusco Maestri, 12: Lambrusco grasparossa, 13: Lagrein, 14: Lambrusco foglia frastagliata, 15: Marzemino, 16: Teroldego, 17–22: wild vines 1–6. (From Scienza, A., Failla, O., Anzani, R., Mattivi, F., Villa, P. L., Gianazza, E., Tedesco, G. and Benetti, U., [1990a]. *Vignevini*, 17(9):25–36. With permission.)

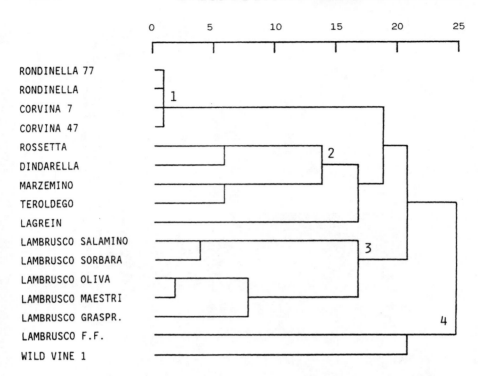

Figure 9. Dendrogram showing the results of cluster analysis carried out on 16 grapevine varieties and 1 wild vine. Cluster analysis was obtained with 53 variables of the electrophoretic pattern of seed protein (see Figure 8). (From Scienza, A., Failla, O., Anzani, R., Mattivi, F., Villa, P. L., Gianazza, E., Tedesco, G. and Benetti, U., [1990a]. *Vignevini,* 17(9):25–36. With permission.)

REFERENCES

Anzani, R., Failla, O., Scienza, A., Campostrini, F. (1990). Wild grapevine (*Vitis vinifera* var. *silvestris*) in Italy: distribution, characteristics and germplasm preservation — 1989 report. *Vitis,* Special Issue, 97–112.

Consiglio Nazionale delle Richerche (1988). Elenco delle cultivar di fruttiferi reperite in Italia. Firenze, Italy.

Levadoux, L. (1956). Les populations sauvages et cultivées de *"Vitis vinifera* L.". *Ann. Amelioration Plantes,* I:59–118.

Mattivi, F., Scienza, A., Failla, O., Villa, P. L., Anzani, R., Tedesco, G., Gianazza, E., and Righetti, P. G. (1990). *Vitis vinifera* — chemotaxonomic approach: anthocyanins in the skin. *Vitis,* Special Issue, 119–133.

Office International de la Vigne et Du Vin (1985). Descriptors list for grape vine varieties and *Vitis* species. Paris.

Rives, M. (1971). Ampélographie. *Sciences et Techniques de la Vigne*, J. Ribereau-Gayon and E. Peynaud (Eds.), Vol. 1, Dunod, Paris, 131–170.

Scienza, A., Protti, A., Conca, E., and Romano, F. (1986). Diffusione e caratteristiche della *Vitis vinifera silvestris* Gmelin in Italia. *Vignevini*, Suppl. 12:86–94.

Scienza, A., Failla, O., and Anzani, R. (1989). Wild grapevine (*Vitis vinifera* L. *silvestris* Gmel.* in Italy: diffusion, characteristics and germplasm preservation. Atti Convegno Int. Symp. Hortic. Germplasm, Cultivated and Wild, Part I, Fruit Trees. Jan. 1988, Beijing, 378–384.

Scienza, A., Failla, O., Anzani, R., Mattivi, F., Villa, P. L., Gianazza, E., Tedesco, G. and Benetti, U. (1990a). Le possibili relazioni tra il *Lambrusco a foglia frastagliata,* alcuni vitigni coltivati e le viti selvatiche del basso Trentino. *Vignevini,* 17(9):25–36.

Scienza, A., Anzani, R., Failla, O., Mattivi, F., Villa, P. L., Tedesco, G., Righetti, P. G., Ettori, C., Magenes, S., and Gianazza, E. (1990b). *Vitis vinifera* — a chemotaxonomic approach: seed storage protein. *Vitis,* Special Issue, 24–28.

Sereni, E. (1986). Per la storia della più antiche tecniche e della nomenclatura della vite e del vino in Italia. *Terra Nuova e Buoi Rossi ed Altri Saggi per una Storia dell'Agricoltura Europea.* Einaudi, Torino, 100–214.

ABSTRACT# 19 A Geographical Information Approach for Stratifying Tropical Latin America to Identify Research Problems and Opportunities in Sustainable Agriculture

P. G. JONES
D. M. ROBISON
S. E. CARTER

Abstract: *Over the last 12 years a database of climate, soils, and crop distribution has been assembled for Latin America. Recently, socioeconomic variables such as access and population density and environmental variables such as the location of national parks, biological reserves, and Indian reserves have been added. Formerly, this information was used primarily to make decisions on commodity research. Given the increasing awareness of long-term agroecological and socioeconomic problems, this database was used to systematize the search for effective, specific courses of research into more sustainable agriculture. Given the premise that agroecological problems and solution vary with both the physical and social environments, the approach was divided into phases. Phase I divided the continent into 128 classes in simple climatic and edaphic terms. The resulting classes were then overlaid with rural population density, rural income per capita, access and location of protected areas. Based on predetermined criteria, a short list of six environmental classes was chosen. Phase II involved a systematic assessment of actual land use in each subzone of the six selected classes. Subzones with similar environments and land uses were grouped in agroecological clusters. These in turn were evaluated for relevance to current and future CIAT research. By this method it was possible to quantify predetermined aspects of sustainability problems based on both environmental and social variables. This formed an immediate basis for deciding between research problems. However, for the long term, it allows systematic comparison between the problems or areas that have been researched and other areas with similar environment or land use problems.*

INTRODUCTION

In the last few decades, international agricultural research centers such as the Centro Internacional de Agricultura Tropical (CIAT) have had clear mandates to attempt to increase total food production to offset a growing population and urban poverty. However, there is a growing consensus that rural poverty and other social problems in tropical countries cannot be solved solely by producing more food. Solutions must include technology that produces food in a manner that protects the natural resource base and is compatible with the given social conditions. Though total food production has increased, other problems have largely been ignored or ill-addressed in the past by mainstream agricultural research. These are vaguely referred to as 'sustainability' problems, and are known to be influenced by both socioeconomic and environmental factors (Douglas, 1984). That is to say, such problems result not only from the nature of the resources but also the given land use and the social factors that drive them. The nonresolution or aggravation of these problems have long-term implications for social welfare, environmental quality, and food production itself.

A problem that CIAT faces in attempting to broaden its research is that it operates in a wide range of environments, both physical and social. For example, though different areas nominally might suffer erosion, the causes and effects differ considerably from country to country and from ecosystem to ecosystem. This supposed site specificity has been seen as an obstacle which impedes technological solutions to problems at a scale larger than that of the individual crop. It would seem that site-specific complexity would preclude an international approach. However, over the past eight years the CIAT Agroecological Studies Unit (AEU) has been conducting crop-specific agroecological analysis in a variety of environments. Fieldwork in similar ecosystems, with similar land use, but in different countries has led to the hypothesis that where climate, soils, and land use are similar, the types of problems tend to be similar. The method described below summarizes one attempt to classify climate, soils, and agricultural land use to spatially identify the most important sustainability problems facing agriculture in Latin America.

Method

The approach taken by the AEU was to classify the continental area in a two-phase process. In Phase I, all of Latin America and the Caribbean were mapped in broad environmental classes. Then, based on predetermined criteria, a short list of environmental classes was chosen. Phase II was the systematic description of actual land use in the selected environmental classes. The most important agroecological clusters (areas with similar environments and land use patterns) and their respective problems were then evaluated for relevance to CIAT's current and future research.

PHASE I

The scope of this first phase was extensive. It included all of Latin America in which CIAT could support a reasonable role in natural resource management. This forced us to make certain assumptions and establish certain criteria for the environmental classification. First, it had to be simple enough to be mapped using available data. Second, it had to be consistent with the data from which it was drawn. Third, it had to reflect the environmental requirements of actual or potential commodity crops for a center of tropical agriculture.

The AEU has detailed data for parts of the continent, but as this scope was broader we opted for more general information consistent across the continent. As the climate database is the most complete, the first step was to classify climate and discard logistically unfeasible classes, thus reducing the total area under consideration.

The Metgrid files used were an interpolation from the climate database, developed in the AEU, which contains mean monthly information from over 7000 stations across Latin America. The interpolation used as a basis the 10' grid of a digital terrain model (NOAA, 1984) and the central pixel from a raster version of the FAO Soil Map of the World provided by the Global Resources Information Database of the United Nations Environmental Program's Global Environmental Monitoring System (UNEP/GEMS/GRID, 1988). From these files we constructed a point quadrat approximation of rainfall, temperature, soils, and elevation for the continent at a spatial resolution of approximately 18.5 km.

Interpolation of the climate data was done by weighted inverse-squared distance from the nearest four stations in the database, corrected for altitude to the NOAA elevation using a standard tropical atmosphere lapse rate model based on data from Riehl (1979). The spatial spread of climate stations is highly variable but tends to be more dense in areas where there is a high variation in altitude and slope and where a majority of the population are often found.

Five environmental criteria were chosen, based on the collective experience of CIAT crop scientists:

Season Length

This was calculated as the number of wet months where rainfall exceeds 60% of potential evapotranspiration, calculated by the method of Linacre (1977).

1.	Humid	10–12	Wet months
2.	Seasonal Wet	6–9	Wet months
3.	Seasonal Dry	3–5	Wet months
4.	Arid	0–2	Wet months

The truly arid classes (two or less wet months) were excluded at this step because CIAT has had relatively little experience with the rain-fed crops or natural resources in these areas.

Temperature During the Growing Season

The growing season was defined as that season with wet months as defined above. The cutoffs were as follows:

1. Lowland tropics   Temperatures greater than 23.5°C
2. Mid-altitude   18 to 23.5°C
3. Highlands   13 to 18°C
4. Cold   Less than 13°C

These temperature cutoffs were based on commonly accepted figures that have proven useful for classifying CIAT's crops in the past. The cold areas were excluded at this step because they represent an area in which CIAT has not worked, and in which other international organizations have a comparative advantage.

Annual Temperature Range

To distinguish between tropical and subtropical areas, we set the annual temperature range cutoff at 10°C.

1. Tropical   Less than 10°C annual range
2. Subtropical   More than 10°C annual range

Diurnal Temperature Range

An additional variable was added to distinguish areas with large diurnal temperature ranges from those with small ranges, based on the experience of the AEU with classifiers that are important for plant growth. This was to differentiate between continental climates and maritime climates, but does not indicate relative distance from the sea in South America, given that the Amazon Basin has an oceanic influence on climate.

1. Maritime   Less than 10°C mean diurnal range
2. Continental   Greater than 10°C mean diurnal range

Soil Acidity

One simple soil variable was used to divided soils into those likely to have serious acidity problems, and those that are unlikely to have such problems. A commonly used cutoff for tropical soils is the pH of 5.5 (Landon, 1984). Below this level the chemistry of many elements changes significantly in terms of toxicity and deficiency. Therefore there were two additional qualifiers:

1. Acid soils   pH less than 5.5
2. Less acid and neutral soils   pH above 5.5

Summary of the Classification System

These variables in theory provided for 128 possible environmental classes. On the one hand, this was an unmanageable number of environmental classes. On the other hand, conditions within each class vary considerably. By eliminating the very dry and very cold areas, the theoretically possible number was reduced to 72 classes. Of these, 9 combinations did not exist in reality, and a further 12 were discarded because either they were too small for consideration, or they were cool subtropical areas with a strong frost risk precluding crops within CIAT's experience.

Stratification

The next step was to stratify these environmental classes in terms of their relevance for future CIAT work. Three broad criteria for choosing environmental classes were given:

1. That the classes be significant for positively affecting rural poverty ("equity").
2. That the classes be important for positively affecting sustainable use of natural resources ("environment").
3. That the classes have potential for increasing food production, thereby benefiting the urban poor ("growth").

To make the stratification possible using these criteria, four independent types of information were combined with the environmental classes using the image overlaying capacity of a geographical analysis package, IDRISI (Eastman, 1988):

Access — As the relative area of a class might be a criterion for choosing between classes, the estimate used was a calculation of the area that is accessible with current infrastructure. Our method was to include the area within each class that was within 30 km of either side of an all-weather road, navigable river, or seacoast. All-weather roads were digitized for each country. For Brazil, this meant digitizing the entire 1989 road atlas. The 60-km corridor along each road is a generous estimate for the increase in access that might occur over the next few years. This analysis can be extended to future development of infrastructure in more detailed studies. For many of the 51 classes this exercise did not reduce effective area by much. However, for the humid and seasonally moist classes it excluded areas such as the Darien Straits, upper Rio Negro, and mid Xingú, which are truly inaccessible, but not legally protected (Figure 1).

Legally restricted areas — The areas in each country in Latin America that are presented legally restricted from conventional agricultural development were digitized from available maps collected by the AEU. These are mostly national parks, forest reserves, Indian reservations, ecological preserves, or protected catchment areas. Some countries report no such areas and in others the protection is only on paper. However, these areas represent a significant proportion of some classes, therefore we excluded them from our calculation of potential agricultural area of an environmental class (Table 1).

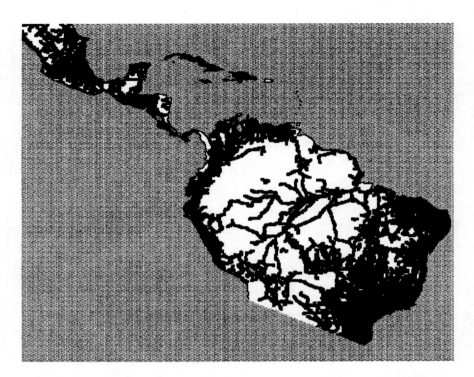

Figure 1. All-weather access in Latin America and the Caribbean. The areas in white are more than 30 km from an all-weather road or navigable river.

Rural population density — Both rural and urban population are unevenly distributed in Latin America. We felt it was fundamental to know the absolute size and relative distribution of the rural population in each environmental class. The nature of most problems and opportunities in agriculture are related to population density and associated infrastructure. As a first approximation we digitized a population map that was transposed from a published population map (Times Atlas, 1985). The actual population represented by this map was calculated by computer and a new map plotted to represent the 1986 rural population. This information was overlaid on the map of environmental classes to provide an estimate of rural and urban population in each class.

Rural income per capita — We included this variable as a crude measure of the magnitude of rural poverty at the country level or, in Brazil, at the state level. Despite its generality, even within Brazil the rural income per capita (PCI) varied from around $150 U.S. (Maranhao and Piaui) to over $2000 U.S. (Mato Grosso do Sul) (IBGE, 1984; World Bank, 1987).

Results

The above socioeconomic information was overlaid onto the map of environmental classes. To achieve a crude assessment of an equity index the mean rural

Table 1 The Effect of Subtracting Legally Protected and/or Inaccessible Land From the Area of an Environmental Class

Class	Description[a]	Rural population	Urban population	Total area (km²)	Number of countries	Accessible area not protected (km²)	Total area not protected (km²)
2	T L S M A	7,462,384	12,830,741	2,800,366	24	810,689	2,431,409
5	T L S C A	4,496,741	8,037,021	1,576,880	18	484,108	1,433,703
17	T M S C A	7,133,114	23,632,759	912,817	18	615,922	846,215
8	T L S M W	5,860,458	8,995,565	540,488	23	303,174	493,803
12	T L D C W	4,704,845	10,728,149	830,303	13	375,999	708,777
9	T L D M W	6,264,550	11,475,161	398,355	12	341,225	391,260
11	T L S C W	4,577,921	8,229,079	390,481	17	180,864	344,035
6	T L D C A	3,471,035	8,324,097	879,678	12	530,767	784,066
1	T L H M A	2,234,896	2,798,926	1,624,899	18	325,642	1,157,602
3	T L D M A	4,122,772	7,077,859	557,513	13	426,590	514,077

[a] T = Tropical, H = Higher altitude, C = Continental, S = Subtropical, H = Humid, M = Maritime, L = Lowland, S = Seasonally wet, A = Acid soils, M = Midaltitude, D = Seasonally dry, W = Weakly acid soils.

income was calculated in IDRISI for each class. The importance of an environmental class for poverty issues increases with the number of people involved, but it decreases as rural income rises. We therefore divided the total rual population in each class by the mean rural income to obtain an index which increased with increasing rural population and/or with increasing poverty. Table 2 shows the classes that ranked the highest for equity.

A subjective productivity index was constructed to rank the environment classes in terms of potential economic impact or growth. This index had values from 1 to 7 per unit area, supposedly proportional to potential biological productivity, and the calculations are shown in Table 3. The potential growth index was calculated by multiplying the area of accessible, legally available land by its productivity index (Table 4).

An effort to rank classes in terms of environmental degradation or risk was more complex, even at this scale, because of the very different types of degradation that exist. An important type of degradation results from nutrient depletion and erosion through insufficient inputs or decreasing fallow time. We have made the assumption that this will occur most frequently in settled areas, but far from markets where there is less incentive to use inputs. The index we used was the area of each class with moderate to low population density (2 to 20/km^2) divided by rural income. Table 5 shows the classes ordered by this index.

A second form of degradation results from ill-conceived intensification, such as excessive agrochemical use. Areas of high risk to these problems will be the higher population areas within each class with easy access to markets, and hence, purchased inputs. The top five classes, by this index, are indicated in the summary shown in Table 6.

A third type of degradation occurs when virgin land is converted to extractive agriculture. Areas with relatively untouched native vegetation, be it forest, savanna, or other types, are likely to be those with low rural populations. Another ranking was made of the area of each class with a population of less than 2/km^2. This can be interpreted either as areas available for expansion of agriculture, or as native vegetation for protection. The top five classes are indicated in Table 6.

Conclusion Phase I

A summary table was calculated which included all the environmental classes that figured in the top five of the different rankings: one for equity, one for growth, and three for sustainability (Table 6). An additional column indicates whether or not the class was in the top five in terms of CIAT's current crop research responsibilities. Within our described method and the criteria we were given, the most relevant classes were 2, 17, 8, 9, and 12. A surprise finding was the importance of class 2 for all the criteria. As a class that is mainly seasonal moist forest, one would not expect it to rank highly in terms of rural poverty. In effect, it has a high total population, mainly along coastal areas, and the rural per capita income is very low, suggesting a large poverty problem. It ranks high

Table 2 Environment Classes Ordered By Rural Poverty Index

Class	Description[a]	Rural population	Rural poverty index	Urban population	Rural pop/km²	Rural PCI mean	Rural PCI std	Number of countries
2	T L S M A	7,462,384	16,480	12,830,741	3	453	298	24
9	T L D M W	6,264,550	11,988	11,475,161	16	523	492	12
8	T L S M W	5,860,458	9,304	8,995,565	12	630	487	23
3	T L D M A	4,122,772	7,619	7,077,859	8	541	460	13
17	T M S C A	7,133,114	6,912	23,632,759	8	1,032	547	18
21	T M D M W	2,544,063	6,674	4,134,194	18	381	170	10
5	T L D C A	4,496,741	6,663	8,037,021	3	675	588	18
14	T M S M A	4,810,238	6,553	13,620,092	14	734	438	21
1	T L H M A	2,234,896	5,677	2,798,926	2	394	111	18
11	T L S C W	4,577,921	5,396	8,229,079	13	848	708	17

[a] T = Tropical, H = Higher altitude, C = Continental, S = Subtropical, H = Humid, M = Maritime, L = Lowland, S = Seasonally wet, A = Acid soils, M = Midaltitude, D = Seasonally dry, W = Weakly acid soils.

Table 3 Relative Productivity Per Unit Area Calculations

		Dry season (months)		
		<2	3–6	7–9
Temperature	Lowland	3	4	2
	Medium	4	4	2
	Highland	4	3	1

Points were first determined for dry season temperature pairs. Then, 2 points were added for nonacid soils and 1 point for subtropical areas.

in environmental concerns because it contains much of the seasonal forest margin in Central America and the Amazon Basin.

A group of economists at CIAT used our extracted data to conduct sensitivity analysis to check for biases towards variables such as class area or population. They used five different scenarios with different factors and weights, independent of area, and essentially the same classes emerged, as is shown in Table 7 (Sanint and Janssen, 1990).

PHASE II — THE DETERMINATION OF LAND USE CLUSTERS

The selection of environmental classes within which to concentrate does not suffice to identify and characterize relevant research problems. Problems with the sustainable management of land resources depend as much on the nature of the land use as on the nature of the resources. The purpose of Phase II, therefore, was to systematize the actual land use in the selected environmental classes. The most prominent combinations of land use and environment were then identified. The nature of problems resulting from the respective land uses, and their relative importance is the kind of information needed by CIAT to plan its research at this scale.

Method

The approach used was to map each contiguous area of a selected environmental class (referred to as a subzone) and determine a number of variables relating to its actual land use. A cutoff size of 600 km^2 reduced the number of subzones in the selected classes from over 500 to just over 300, yet accounted for over 98% of the area.

The percent area in three topographic slope classes (0–8%, 8–30%, >30%) was estimated from medium scale topographic maps. Soil depth, predominant texture, drainage, and any obvious chemical or physical problems were noted from semidetailed soil maps. The number of months with over 200 mm precipitation was calculated from the CIAT database. In those countries where a relatively recent agricultural census was available, percent area under annual

Table 4 Environment Classes Ordered by Production Potential Index. This Was Calculated By Multiplying the Relative Productivity Index By the Accessible Area of Each Class

Class	Description[a]	Subjtv. prod. index	Sum. prod. index	Rural population	Rural pop/km²	Number of countries	Accessible area (km²)
2	T L S M A	4	3,242,757	7,462,384	3	24	810,689
5	T L S C A	4	1,936,433	4,496,741	3	18	484,108
17	T M S C A	3	1,847,765	7,133,114	8	18	615,922
8	T L S M W	6	1,819,042	5,860,458	12	23	303,174
12	T L D C W	4	1,503,994	4,704,845	7	13	375,999
9	T L D M W	4	1,364,902	6,264,550	16	13	341,225
11	T L S C W	6	1,085,185	4,577,921	13	17	180,864
6	T L D C A	2	1,061,534	3,471,035	4	12	530,767
1	T L H M A	3	976,925	2,234,896	2	18	325,642
3	T L D M A	2	853,181	4,122,772	8	13	426,590

[a] T = Tropical, H = Higher altitude, C = Continental, S = Subtropical, H = Humid, M = Maritime, L = Lowland, S = Seasonally wet, A = Acid soils, M = Midaltitude, D = Seasonally dry, W = Weakly acid soils.

Table 5 Environment Classes Ordered By Nutrient Depletion/Environmental Degradation Index (e.g., Erosion or Nutrient Leaching, Weed Infestation)

Class	Description[a]	Nutrient depletion degradation index	Rural population	Urban population	Rural pop/km²	Rural PCI mean	Number of countries	Accessible area
2	T L S M A	792	7,462,384	12,830,741	3	453	24	810,689
3	T L D M A	517	4,122,772	7,077,859	8	541	13	426,590
9	T L D M W	473	6,264,550	11,475,161	16	523	12	341,225
5	T L D C A	449	4,496,741	8,037,021	3	675	18	484,108
17	T M S C A	427	7,133,114	23,632,759	8	1,032	18	615,922
6	T L D C A	386	3,471,035	8,324,097	4	882	12	530,767
21	T M D M W	308	2,544,063	4,134,194	18	381	10	130,436
18	T M D C A	292	3,379,676	8,204,852	7	826	12	362,535
12	T L D C W	283	4,704,845	10,728,149	7	954	13	375,999
1	T L H M A	235	2,234,896	2,798,926	2	394	18	325,642

[a] T = Tropical, H = Higher altitude, C = Continental, S = Subtropical, H = Humid, M = Maritime, L = Lowland, S = Seasonally wet, A = Acid soils, M = Midaltitude, D = Seasonally dry, W = Weakly acid soils.

Table 6 Summary of Phase I: The Occurrence of Classes in the First 5 Rows of the Subject Rankings (Tables 1, 2, 4 and 5)

| Class | Growth | Equity | Environment | | | |
			Intensification abuse	Protection	Nutrient depletion	CIAT[a] crops
2	*	*	*	*	*	*
17	*	*	*		*	*
5	*			*	*	*
8	*	*	*			
9		*	*		*	
12	*		*	*		
3		*			*	
1				*		
6				*		
18						*

[a] Rice, beans *(Phaseolus vulgaris)*, and cassava.

Table 7 Results of Sensitivity Analysis on the Environmental Classes, Using Five Different Weighting Scenarios

Scenario	1	2	3	4	5
Top 5	8	9	8	2	8
	2	2	2	9	2
	9	8	9	8	9
	17	17	17	17	17
	20	20	11	20	11
Second 5	11	21	20	21	12
	12	11	12	11	20
	21	12	21	12	5
	5	5	5	5	21
	14	34	14	23	14

From Sanint, L. and Janssen, W. (1990). *Methodology for Ranking LAC Environmental Classes*, Centro Internacional de Agricultura Tropical, A.A. 6713, Cali, Colombia. With permission.

cropping, perennial cropping, pasture, forest or fallow was calculated. In other countries, this was estimated from land use maps. Socioeconomic variables were also estimated for each subzone such as population density, urban dependence on agriculture, land distribution, percent of area readily accessible to transport, and relative distance to market.

Once the worksheets had been filled, the topographical, agricultural, and social information was used to determine generic production systems for each of the 300 subzones such as 'extensive cattle grazing' or 'intensive irrigation of annual crops'. It is important to note that virtually all of the subzones had at least two modal production systems practiced by different people within the same subzone, e.g., extensive cattle ranching by large landholders and shifting cultivation by small landholders. These interacting production systems together were termed land use patterns, and a land use pattern was assigned to each of the 300 subzones.

Table 8 Principal Land Use Patterns Identified

Extensive cattle/shifting cultivation/forest	XC-SC-F
Extensive cattle/mechanized annual crops/shifting cultivation	XC-MA-SC
Hillside cattle/coffee/horticulture	HC-CO-HO
Hillside cattle/coffee/shifting cultivation	HC-CO-SC
Intensive sugar cane/intensive cattle/mechanized annual crops	IS-IC-MA
Intensive irrigated crops/extensive cattle	II-XC
Rubber and Brazil nut extraction/forest	RN-F
Traditional riverine systems on flooded land	TR-F
Extensive goat grazing	XG
Mechanized coffee/mechanized annual crops/intensive cattle	MC-MA-IC
Extensive cattle/forest	XC-F
Extensive cattle/mechanized annual crop/forest	XC-MA-F
Small scale sugar cane and annual cropping	SS-SA
Intensive irrigation/medium scale annuals	II-MM
Medium scale mechanized annual/medium scale cattle	MM-IC
Extensive cattle on poorly drained soils	XCP
Shifting cultivation/managed forest/small scale cattle	SG-SC-BA
Small scale cattle/shifting cultivation/commercial bananas	SG-SC-BA

The order of the abbreviation does not always represent the relative predominance of the individual systems.

Table 8 shows all of the land use patterns identified. Instances of repeating land use patterns within an environment class was termed an agroecological cluster. Table 9 illustrates some differences between three agroecological clusters within one environmental class. Comparison of Figure 2 and Table 8 shows that just over one third of the potential combinations of land use patterns and the seven environmental classes exist. Some land use patterns are not significant in some environmental classes.

Land use patterns appear to be repeated to the extent that geographically separate subzones have similar physical and human environments. They are expressions of the relationships between the landscape and the natural environment, and the social and economic conditions under which agriculture is practiced. For example, where neutral soils, long growing seasons, good access, and close markets combined, the predominant land use in Latin America was intensive sugar cane and intensive cattle. Where mid-altitude temperatures, good access, acid soils, and steep slope were found together, the land use was predominantly coffee and intensive cattle with some horticulture. A third example was where poor access, large distance from market, and natural forest vegetation occurred together, the land use was predominantly shifting cultivation and extensive cattle grazing (e.g., the forest frontier).

Not only do the individual production systems interact with the environment, but different systems within an area also interact and compete with each other for resources, thus forming part of the overall environment. From the knowledge gained in describing the agriculture in each subzone, we may assume that the types of problems faced (environmental, social, and economic) are similar for different subzones with the same land use patterns and environment. Amongst those cells which are recorded (Table 10), it is relatively straightforward to

Table 9 Example of Some Topography and Land Use Variables For Three Different Agroecological Clusters Within One Environmental Class, 17 — Tropical, Midaltitude, Seasonally Wet, Continental, Acid Soil

	Total area (000) km²	No. of countries	Slope (<8%)	Classes (8–30%)	Land use (% area)			Access	Properties (%>10 ha)
					Annual crops	Perennial crops	Pastures		
MC-MA-IC[a]	197	1	15	67	13	6	64	100%	26
HC-CO-HO[b]	98	10	8.5	43.5	14	18	42	63%	65
HX-MA[c]	311	2	45	33	13	0.01	43	76%	2

[a] Mechanized coffee, mechanized annual crops, intensive cattle.
[b] Hillside cattle, coffee, horticulture.
[c] Extensive cattle, mechanized annuals.

LAND USE PATTERN / ENVIRONMENT CLASS

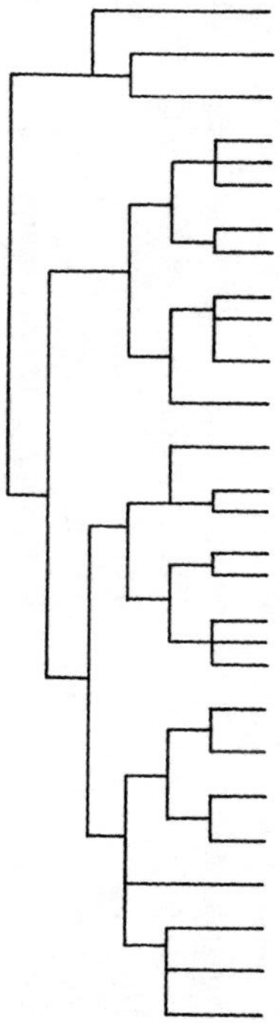

LAND USE PATTERN	2	5	8	9	11	17	20
UNUSED FOREST LANDS	1.68	0.17					
RUBBER - NUTS	2.58	1.88			0.77		
PLUVIAL & VARSEA SYSTEMS	6.11	0.18	1.62		1.77		0.46
INTENSIVE CANE POOR LANDS	7.51		15.57		1.73		
INTENSIVE CANE GOOD LANDS			1.27	6.34	0.24		0.72
INTENSIVE IRRIGATION							
BRASIL MECHANIZED COFFEE AREAS	3.17	4.09				19.73	
MECHANIZED MEDIUM SCALE			1.24		0.23	0.18	
CERRADOS TYPE PASTURES MECH CROPPING	11.27	29.22				31.07	
POOR LOWLAND PASTURES MECH CROPPING	3.35	1.06					
LOWLAND EXTENSIVE GRAZING POOR SOILS							
GOOD LOWLAND PASTURES MECH CROPPING	4.49	1.60	3.39		2.10		0.52
POORLY DRAINED PASTURES			0.11		8.42	0.35	
GOOD LOWLAND PASTURES ALONE					0.54		
HIGHLAND PASTURES ALONE	37.42	7.26	4.98				
POOR LOWLAND PASTURES MANUAL CROPPING				5.10	1.73		
GOOD LOWLAND PASTURES MANUAL CROPPING			0.79	13.92			
DRY LOWLAND PASTURES MAN/MECH CROPS				4.44			
DRY LOWLAND PASTURES MANUAL CROPPING							
GOAT GRAZING						3.02	0.22
HILLSIDES CATTLE COFFEE POOR SOIL					0.08	6.78	3.51
HILLSIDES GRAZING SHIFT CULT POOR SOIL							
HILLSIDES CATTLE COFFEE GOOD SOIL							2.89
HILLSIDES GRAZING SHIFT CULT GOOD SOIL	2.05		0.15	1.62	0.17		
LOWLAND CATTLE COFFEE			0.84	0.56			0.13
SHIFTING CULTIVATION	0.26		0.05	1.47	0.23	0.42	0.26
SMALL SCALE CANE & MANUAL CULTIVATION	1.11	2.92	0.26	1.62			
LOWLAND GRAZING. SHIFT CULT ON SLOPE							

Figure 2. Area (millions of hectares) of the main agroecological clusters in the seven environmental classes. The cluster diagram at the left indicates relative resemblance of the different land use patterns.

Table 10 Summary of the Most Important Agroecological Clusters

Land use pattern	Selected environmental classes						Other classes
	2	5	8	9	17	20	
Extensive grazing/manual cropping	XXXX	XX	XX	XXX			XX
Extensive grazing/ mechanized cropping	XX	XXX			XXXX		XX
Hillside grazing/coffee/ shifting cultivation			X	X	XXXX	XX	XX
Mechanized coffee/ pasture/mech. crops					XXXX		X

identify the agroecological clusters which have the greater relative importance in terms of area and population.

Application

Figure 2 and Table 10 provide information, for CIAT or any potential user, from which to make decisions about the relative importance of different land uses and their problems. At CIAT, the former criteria were used to indicate areas where it would be logical to begin research on sustainable agriculture and its relationship with environmental and socioeconomic conditions. When sorted by predominant land use patterns, a series of groupings appeared which seemed to be logical. These were inspected and clustered according to a consensus of subjective estimates of similarity among those working in the AEU. Since much of the information was nonnumeric and not ordered, this was considered more appropriate than a numeric clustering algorithm. Figure 2 shows the areas and population, respectively for these land use pattern groups within the environmental classes selected in Phase I.

Once the AEU had provided the basic information on the different agroecological clusters, a multidisciplinary group of CIAT scientists selected the three most relevant (CIAT, 1991).

The first was termed the seasonal forest margin, and consists of lowland areas of manual cultivation and extensive grazing, with a seasonally wet climate. Continental and maritime instances of this land use pattern were combined to define the focuses for research. The areas in question have very large expanses of degraded pasture, the rehabilitation of which has long been a concern of CIAT's Tropical Pastures Programme. A significant amount of upland rice and cassava also occurs, particularly in Brazil. Current land use is not sustainable, in part contributing to further deforestation.

The second agroecological cluster was composed of the seasonally wet hillsides of the northern and central Andes, Central America, and the Caribbean. Intensive coffee production and cultivation of annuals, in association with extensive pastures, is very important. Cassava and beans are important staples in this area, and cattle are common as a source of milk, meat, and cash on both small and large farms. Deforestation, erosion, agrochemical abuse, and fragmentation are among the problems encountered.

The third agroecological cluster chosen is that which contains extensive grazing with or without large-scale mechanized agriculture, on the natural savannas of the Llanos (Colombia and Venezuela) and Cerrados (Brazil). Lowland and midaltitude, seasonally wet environments have been combined to define the area for research to focus on. Research at CIAT into the intensification of these extensive grazing systems through the incorporation of annual crop rotations has become increasingly important over the last few years.

The methodology and data were employed initially to select those agroecological clusters which were most important for a given agricultural activity or type of research. However, in the future, they should primarily promote an understanding of the similarities and differences between individual agroecosystems. Similar land use patterns are found in dissimilar environments and vice versa, but the degree of similarity can be assessed from Table 10, with a knowledge of the environmental classes. For an institution such as CIAT, which wishes to generate new agricultural technology, this is critical. Innovations which modify land use systems may have applicability across different environments. An understanding of environmental conditions can provide a rational frame for evaluating innovations in areas which are environmentally distinct from those where adoption has occurred. Similarly, within a single environment class, it is vital that researchers understand land use patterns if they are to increase their understanding of farmers' needs for new technology.

CONCLUSION

It was impossible to consider all land use problems in the entire continent. However, by the above method the AEU was able to systematically identify and quantify widespread specific land use problems. Our approach offers a distinct advantage over more subjective attempts to identify areas in which to conduct research, whether for agricultural development, environmental protection, or the conflict between these two goals. Agroecological zonification based on physiological requirements of single crops (FAO, 1978) alone cannot help to understand sustainability problems. Similarly, studies to determine the ideal or potential uses of land, without studying the limitations imposed by actual land use, are of limited utility. An approach that includes both environmental and social variables provides a means to select locations and agrarian problems systematically, and hence to relate the results of research rationally to other related places or problems.

By tentatively defining a series of relationships between man's activities and environmental conditions, expressed as agroecological clusters, the work has provided the basis for systematic study of agricultural systems and their environmental consequences. What are needed now are comparative studies of the interactions between the different production systems which make up the land use patterns. This is vital if we are to understand the way in which the actions of certain groups within agrarian societies, the intended beneficiaries, affect productivity and the environment.

ACKNOWLEDGMENTS

The authors wish to thank Humberto Becerra and Otoniel Madrid for the buildup and maintenance of the climatic database through the years. We are also grateful to Silvia Elena Castaño, Ligia García, and Elizabeth Barona for digitizing, reference, and cartographic help; to Yúviza Barona and Marta Lucia Gómez for secretarial help; and to Mauricio Rincón for data transformation and verification and for help in describing the subzones.

REFERENCES

CIAT (1991). Centro Internacional de Agricultura Tropical (CIAT) in the 1990s and Beyond: A Strategic Plan, (2 parts). Cali, Colombia. Part 2 (Suppl.), 85–125.

Douglas, G. K. (Ed.) (1984). *Agricultural Sustainability in a Changing World Order*, Westview Press, Boulder, CO, 282.

Eastman, R. J. (1988). IDRISI. A Grid Based Geographic Analysis System. Manual to version 2.24. Clark University, Worcester, MA, 361.

FAO (1978). Report on the Agro-Ecological Zones Project (World Soil Resources Report 48). Vol. 1 Methodology and Results for Africa. Food and Agriculture Organization, Roma.

IBGE (1984). IX Recenseamento Geral do Brasil 1980: Censo Agropecuario. Fundacao Instituto Brasileiro de Geografia e Estadistica. Rio de Janeiro, 494.

Landon, J. R. (Ed.) (1984). *Booker Tropical Soil Manual*. Longman, New York, 455.

Linacre, E. J. (1977). A simple formula for estimating evaporation rates in various climates using temperate data alone. *Agric. Meteorol.*, 18:409–424.

NOAA (1984). TGP-006D Computer Compatible Data Tape. National Oceanic and Atmospheric Administration, Boulder, CO.

Riehl, M. (1979). *Climate and Weather in the Tropics*. Academic Press, London, 67.

Sanint, L. and Janssen, W. (1990). Methodology for Ranking LAC Environmental Classes. Mimeo Planning Document. (Unpublished). Centro Internacional de Agricultura Tropical, A. A. 6713, Cali, Colombia.

Times Atlas (1985). The Times. *Atlas of the World*. Times Books, London.

UNEP/GEMS/GRID (1988). FAO Soils Map of the World at 30 Seconds Resolution. Computer Compatible Data Tape United Nations Environment Program Global Resource Information Database. Geneva.

World Bank (1987). World Development Report 1987. World Development Medication. World Bank, Washington, D.C., 202–207.

20 Ecological Aspects of Production in the Canary Islands: Traditional Agrosystems

V. MARTIN MARTIN
W. RODRIGUEZ BRITO
A. BELLO

Abstract: *Three different areas representative of traditional agricultural systems in the Canary Islands were studied in order to find the ecological basis governing their productivity. Traditional agricultural techniques could bring solutions to problems of great importance in Mediterranean and mountain environments, such as desertification and pollution. These phenomena progress rapidly in island areas of volcanic origin that are fragile and under pressure from tourism, and where the complementary nature of agricultural systems has not been taken into account in land management.*

Aridity and moisture were the basic factors for choosing the agricultural systems studied. Agrosystems in arid environments have been studied in southern Tenerife and Lanzarote. They can be distinguished by the volcanic material that is used to prevent aridity ("jable" and "lapilli", respectively). Agricultural systems in a humid environment are studied in northern Tenerife, which is under the influence of trade winds. The study was carried out by two coordinated teams. The first did the spatial analysis of the agricultural systems in order to find the key processes governing them and the second was concerned with the ecological basis for interpreting these processes. Results have enabled the identification of a series of cultural techniques which are the keys to environmental pest and disease control, combatting the wind, aridity, and erosion, and maintaining soil fertility without pollution. Such measures give rise to agricultural patterns with quality products, environmental conservation, and landscapes of great aesthetic value, as well as encouraging the maintenace of a cultural asset of great interest in searching for solutions to the struggle against aridity and desertification.

INTRODUCTION

The growing importance of research on the ecology of agricultural systems is no accident. Terms such as "limits of economic growth", "sustainable development", or "ecological economics" stress the management of natural resources from standpoints which go beyond those of the mechanics of utility and benefit (World Environmental & Development Commission, 1987). A new relationship between man and space is being generated, in which productivity is no longer the driving force and where land management is being undertaken from multiple disciplines (e.g., anthropology, sociology, biology, and economics).

"Agroecology can best be described as an approach that integrates the ideas and methods of several subjects, rather than as a specific discipline" (Hecht, 1987). Agroecology has introduced a new concept which may contribute to clarifying the man/environment relationships: the agrosystem. "The challenge for agroecology, then, is to find a research approach that consciously reflects the nature of agriculture as the coevolution between culture and environment, both in the past and the present. The concept of the agrosystem can (and should) be expanded, restricted or altered as a response to the dynamic relationship of human cultures and their physical, biological and social environments" (Gliessman, 1990).

This new discipline, agroecology, endeavors to give a scientific explanation to the traditional agricultural systems which have been self-sustained throughout the years without the use of excessive inputs.

A great variety of agrosystems exists in the Canary Islands, despite the limited area. These agrosystems are built around what is known as an internal or subsistence market agriculture, although crops for export outside the Islands are grown. Three representative models of the traditional farming systems in the Canary Islands have been studied with the purpose of establishing the ecological criteria governing their productivity.

The Canary Islands agricultural systems are of interest for the following reasons:

- The traditional farming techniques could contribute to solving important problems in Mediterranean environments, such as desertification, by combating aridness, erosion and wind, or pollution by crop rotation and using multicrops. It should be taken into account that these substrate are totally volcanic.
- Farming systems in the Canary Isles have given rise to quality agriculture, as well as conserving the environment and generating landscapes which, at present, constitute a valuable cultural asset.
- Intense land development is taking place, because of tourist development and demographic pressure, which is causing destruction of traditional agricultural systems.

The interaction of the above three factors requires future solutions which will involve making agriculture and tourism complementary to each other.

PHYSICAL VARIABLES: AN ADVERSE ENVIRONMENT FOR AGRICULTURE

Of volcanic origin, the Canary Island Archipelago is formed by eight inhabited islands and several islets, with a total area of 7511 km^2. Its geographical location determines two physical facts: a subtropical latitude (28° North) close to the Tropic of Cancer, and its proximity to the western face of the continent of Africa. There is scarcely 100 km between the most eastern island and the African coastline, where the most extensive desert in the world, the Sahara, begins (Figure 1).

Nevertheless, the Canary Isles have a climate which is not appropriate to their geographical location. This is due to the influence, on the one hand, of the Azores Depression which forms part of the what is known as a high-pressure, subtropical belt and, on the other hand, the Canary Island Cold Sea Current.

The geographical location, regional atmospheric circulation, and cold sea current explain the general Canary Islands climatic features: characterized by mild average temperatures with a low thermal range (20°C average annual temperature), high relative humidity, total precipitation between 200 and 400 mm/year, and cool NNE winds (the trade winds).

The climate's general characteristics have a high variety of local microclimates which noticeably influence agricultural systems. Such microclimates result from key factors such as the geographical situation of each island, insular orography and altitude, exposure, and arrangement of the relief. The following microclimate elements define the agrosystems studied (Figure 2).

Lanzarote — The island's maximum altitude is less than 700 m (670 m at Peñas del Chache) and therefore, its whole area is below the level of trade wind reversal. This physical fact restricts the important orographic rainfall in the western isles (Marzol Jaen, 1988). There is no mountainous obstacle that can stop the "sea of clouds" (stratocumulus bank of clouds caused by stratification of trade wind air masses). Moreover, together with Fuerteventura, Lanzarote is the nearest island to the continent of Africa. As a result, the annual average rainfall is around 160 mm and never exceeds 300 mm. Rainfall is irregular and does not exceed 47 days annually. Insolation is high (65%), with 130 clear days. The average temperature exceeds 20°C, with strong daylight sunshine (this is the island with the greatest daytime thermal range). The relative humidity is around 70%, but is reduced by the strong insolation. Winds are frequent but not violent (breezes and trade winds).

Tenerife — On the island of Tenerife (with a maximum altitude of 3717 m in the Teide composite volcano), masses of air change their behavior due to the orography. The orientation and thickness of the relief force the cloud masses from the NW and NE to collide along the island's north face, discharging part of their water content into the northern side and then falling with a föhn effect onto the south face (Marzol Jaen, 1988). These phenomena produce a marked differentiation between the island's north and south face. For example, in Las Mercedes — NE facing — 1000 mm of rain are collected per year on average, while at Punta de Rasca — south facing — it does not reach 100 mm per year. As a result, the northern face is characterized by an evenly spread system of rainfall ranging from autumn to spring (400–600 mm annual average), cool temperatures, high relative humidity, and abundant cloudiness. On the leeward

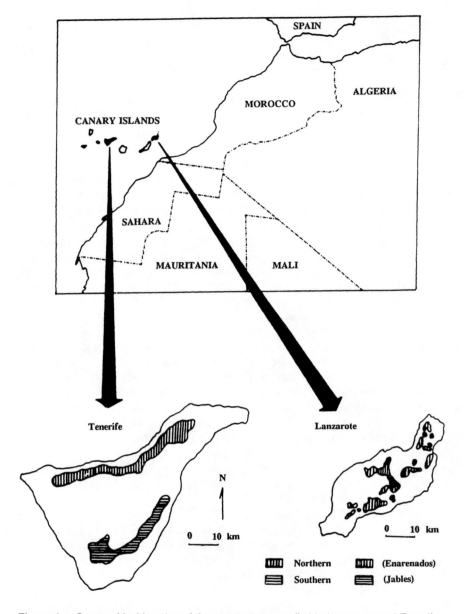

Figure 1. Geographical location of the agrosystems studied in Lanzarote and Tenerife.

face, with many hours of sunshine, there is little cloudiness and an average precipitation of between 100 and 300 mm.

The volcanic nature of the archipegalo, together with its climatic features, are the other physical factors affecting the traditional Canary Isles farming systems. In Lanzarote, a large part of the island is covered with recent volcanic materials,

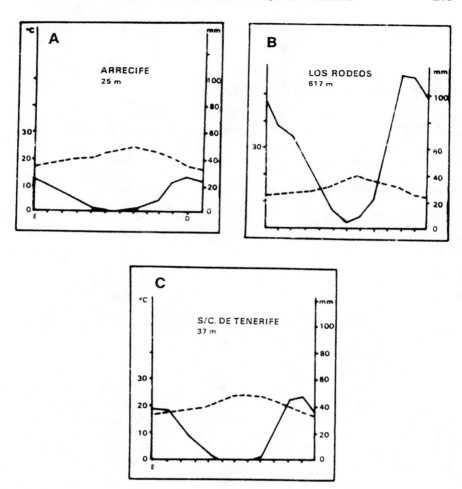

Figure 2. Gaussen's ombrothermal diagram of the three agrosystems in different seasons. (A) Lanzarote; (B) Northern Tenerife; and (C) Southern Tenerife.

with eruptions dating from the 18th and 19th centuries. The result is that a major part of the island surface is covered by "malpais" and lapilli (volcanic cinders). "Jables" or calcareous sands of a marine origin are abundant on this island, formed by remains of shells and seaweed. Their origin lies in the Famara beach area to the north of the island. The predominant NE winds blow the organogenous sands, which the sea leaves on the beach, from Famara to the west coast.

On Tenerife's north face are found the most developed soils on the island (brown/fersialitic) gained for agriculture by felling the thermophilic forest (beech/heath and "laurisilva"). In the south part of Tenerife there are outcrops of a group of pyroclastic materials of a salic nature, known as pumices, which give a certain homogeneity to the landscape.

MAN IN AN ADVERSE ENVIRONMENT

In the adverse environment described above, an agricultural culture has been developed which may be defined as man's struggle against the environment, without damaging the island ecosystem. The development of the man/environment relationship in the three agrosystems studied is described below.

Lanzarote — Modern Agriculture With Craft Techniques

There are two forms of cultivation as far as the nature of the agricultural substrata is concerned: the "jables" (organogenous sands) and the "enarendados" (lapilli of a basic origin).

Jables

Jables cover the argillaceous soil. After plowing the soil, holes are dug ($1/4$ m apart) until clay is reached, dung and chemical fertilizers are thrown in together with the seed, and all is finally covered with jable. Thus sweet potatoes and, to a lesser extent, marrows, sprouts, tomatoes, watermelons, and other crops are grown. Crops are rotated every three years (sweet potatoes/tomatoes/fallow) or two years (sweet potatoes/fallow).

In order to combat the constant NE winds blowing in the island, the crop plots are fenced in with "bardos": mud walls generally mixed with rye straw. The bardos are placed perpendicular and parallel to the predominant winds. Another of their functions is to stop wind-blown sand. Otherwise, the above-ground part of the plants would not only be knocked over but would be buried, since they act as an obstacle (Rodriguez Brito, 1986).

The evident aridity of Lanzarote is attenuated by using jables, since this sand of an organic origin preserves and condenses humidity. Likewise, it favors the growth and development of plants through acting as a heat insulator. An advantage of the jables is that it enables vegetables to be grown which demand water in a rain-fed system.

After passing through a peak period in the 1960s the jables were gradually abandoned. Nowadays, sweet potatoes and watermelons are sown over an area of less than 400 ha.

Enarenados

The enarenados are made up of soils covered with lapilli which have the effect of retaining and preserving subsoil moisture. They are used in two ways:

Natural Enarenados These are found in areas near recent volcanoes where lapilli covers the preexisting plant soil. The fact that lapilli thickness is highly variable makes farming difficult, as a thickness between 0.2 and 2 m is required.

Crops which can use these "enarenados" must have a deep root system enabling them to penetrate through the lapilli layer. The predominant crop, therefore, is usually the vine and, to a lesser extent, fig and other fruit trees.

Farmland is prepared by digging holes from 1 to 1.25 m deep, until the paleosoil is reached. A vine or fruit tree is planted at the bottom — 1 ha can take between 250 and 350 plants. The holes are protected with low stone walls which are placed perpendicular to the predominant wind direction. In this way, farming has been achieved in a highly xerophillous environment, with heavy evaporation and intense sunshine (these crops are located in the island's most arid sector).

Artificial Enarenados Their use is also to preserve soil moisture. The ground is cleared of stones for farming, and a 10- to 15-cm layer of lapilli is placed on it. The soil can keep moisture in even 12 months after the last rainfall, and agricultural yields are thus obtained comparable to those from irrigated land, though precipitation hardly exceeds 100 mm/year (Rodriguez Brito, 1986).

The onion is the predominant crop but melon, watermelon, sweet potatoes, pulses, potatoes, maize, and other cereals also occur. The variety of crops which have gradually adapted to Lanzarote's agrosystem must be related directly to the island's traditional isolation, which has caused the need for a certain self-sufficiency as far as food production is concerned. In addition, soil exhaustion caused by sweet potatoes, watermelon, or melon farming requires rotation with pulse crops and even fallow lands.

Lanzarote is shown as an area where varieties have been selected to adapt to the arid environment. Sweet potatoes, watermelons, and melons are all crops of a tropical origin (Central Africa and Monsoon India), highly thermophilic, and which absorb quite an amount of humidity. Adaptation to the island environment is also demonstrated in some varieties of potatoes and maize (dwarf maize, low in height with a single root system). The onion is the most widespread crop due, in part, to its high productivity, and in part to an exchange of seeds with other islands.

Artificial enarenados play an important role in the fight against erosion, as they allow rainwater to infiltrate, thus preventing runoff.

The problems of enarenados derive from their low average lifetime, from 15 to 20 years, at the end of which they must be renewed since they have mixed with the soil. There are also problems related to tillage, fertilization, and pest treatment.

In conclusion, Lanzarote's agriculture has an exceptional character due to the adverse environmental conditions. The existence of important agriculture is due to man's inventiveness which has solved the disadvantages of the situation. Farming in enarenados and jables enable yields per hectare to be similar to those on irrigated land. This type of farming has proved to be a curb on erosion in an arid environment, where the scarce resources that exist for developing agriculture of quality and landscape creation have been optimized.

Tenerife

The two agrosystems studied are located in the "medianias" strip in the North (400–1000 m) and South (500–1200 m) on the island of Tenerife. Despite the different climate and edaphological characteristics, both systems are complementary on the level of seed exchanges, specialization of crops, or livestock activities. Nevertheless, nowadays this complementary nature is breaking down, constituting one further element in the agricultural crisis of the Canary Isles.

The North of Tenerife: Control of Inputs and Combatting Pollution

The medianias of the north part of Tenerife cannot be defined as a single agricultural system because there exists a great horizontal and vertical variation in the agrosystems, depending on the edaphic substratum and local microclimates. They do however offer the same ecological bases which enrich them: a diversity of crops (rotations and multicrops) for controlling pests and minimizing the consumption of inputs, and the importance of livestock in the struggle against the scarcity of soil and erosion.

The fight to maintain fertility has forced extensive rotation of crops in view of the scarcity of both natural and artificial fertilizers. On the other hand, an agricultural economy where livestock occupies a leading position (beef, and to a lesser extent, goats, pigs, mules, and asses) requires forage crops. In turn, this complementary nature between agriculture and livestock is demonstrated in the contribution of organic fertilizer (dung) from livestock. In one of the most interesting areas, a particular agricultural system in the north of Tenerife (La Esperanza), up to five-year rotations are undertaken. Thus, on the same plot, potatoes, fallow land ("manchón") for two years, cereal (generally wheat), and lupin are the alternatives included in the rotation. During spring, a short-cycle crop, like maize, can be sown.

This long crop rotation system fulfills various functions whose main objective is to maintain fertility without pollution effects with a low input consumption and a complementary arrangement between agriculture and livestock practice. Combating pests in these long rotation systems has been shown in the control of nematodes in potato growing.

There exist other types of more simple rotations distributed throughout the mediania strip in the north of Tenerife: e.g., three-year rotation with tilled fallow land (fallow, wheat, and potatoes) and rotations without fallow land (wheat and potatoes) (Alvarez Alonso, 1976). On occasions when sufficient natural fertilizer is available, the crop can be intensified and three crops of potatoes on the run can be obtained, but afterwards the rotation system is reintroduced.

Another ecological base which contributes towards enriching the organization of the northern agrosystem is multiple cropping, e.g., vines/potatoes, potatoes/

maize/leguminous plants, potatoes/sprouts, beans/maize, potatoes/fruit trees and vines/maize/sprouts. The location of the mixed crops within the area under study varies as a function of the edaphic substratum and microclimates. For example, the vine does not appear in the long La Esperanza rotations which were discussed earlier, because this area is exposed to wind and is excessively wet and rainy. However, farming happens in areas of low medianias (Acentejo District) and in others where the soil is less developed, as is the case of lapilli or the volcanic conglomerates of an intermediate or salic nature (Icod District). Vines may appear as the only crop when conditions are favorable.

The beans/maize association is located in the low medianias (rarely above 700 m) in the whole of the north but mainly in the northwest end (La Culata depression) where there is less cloud and temperatures are mild.

One of the most interesting mixture of crops is the potato/fruit tree, with the latter represented by the chestnut tree. This is a tree suited to the wet conditions of the north of the island, mostly located in the Acentejo District. The importance of this crop association lies, amongst other features, in its providing shade and environmental moisture in the last maturing stage of the potato. These phenomena have been seen when warm air masses arrive from the neighboring continent of Africa.

The exchange of seeds between farmers of both zones and adaptation of different plant varieties are other elements distinguishes this agrosystem. This is shown by the traditional seed exchange between areas in the north of Tenerife itself (both vertically and horizontally) or from the north with the south and vice versa. The adaptation of varieties is demonstrated in potato growing, with local varieties like "bonita", "negra", "colorada", "rosada", "melonera", "venezolana", "liria" and others of a British origin, e.g., "King Edward", "Up-to-date", and "Cara".

Erosion, wind and scarcity of soil are the adverse elements for farming in this area. To reclaim agricultural land it has been necessary to create a system of terracing in steeply sloping hill sides. At the same time, the terrace system involves an obvious curb on erosion processes (particularly those due to water erosion). Growing quickset hedges like the "tagasaste" *(Citisus proliferus)* on the edge of the plots has not only acted as fodder for livestock but has also turned out to be an effective means of attenuating erosion (their roots act as a soil fixer), combating the wind, and giving the soil fertility. Quickset hedges are of a variety of plants, e.g., trees, agaves, prickly pear, broom, and heather.

The main problems currently within this agricultural system are both internal and external. The former include the agricultural crisis of the Canarian domestic market, where only the vine and potatoes generate sufficient profit for the farmer. The latter is caused by land development taking place throughout this mediania strip (Santa Cruz/La Laguna metropolitan area stretching from the Puerto de la Cruz tourist center) and the growing drift of population towards the service sector. Both phenomena are generating part-time farming around potato and vine growing (Burriel de Orueta, 1981), (Figure 3).

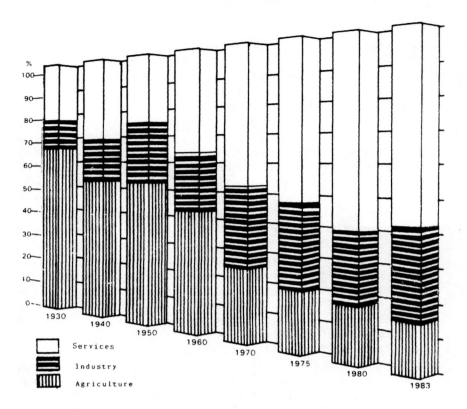

Figure 3. Evolution of the working population in the three largest Canary Island sectors (1930–1983).

South Tenerife: Adaptation of an Undiversified Agricultural Systems to an Arid Environment

The agricultural systems in southern Tenerife contrast strongly to those in the north. In the north, the rotation system and the associated crop are most important, but in the south it is the adaptation to an arid environment where livestock was possibly traditionally of greater importance than farming. The absence of rotation systems and associated crops is therefore one of the main problems to be solved in the southern agrosystem.

The major environmental elements that have been developed in the agriculture of this area are the pumice "enarenados", "gavias", and "nateros" systems.

Pumice Cultivation Two subtypes are distinguished: direct growing on pumice and argillaceous soil covered with pumice (here also called jables).

The first subtype is characterized by growing on disturbed pumice (fallen acid-origin pyroclasts). Their abundance and facility for working them have facilitated a very peculiar landscape since both the habitat and the growing soils are located in the places of greatest abundance of these volcanic materials. The scarcity of developed soils in this part of the island restricted settlement until the 19th century, when pumice began to be used in agriculture. Therefore, it became, in the 30s and 40s of the present century the most prosperous district in the south of the island. The pumice allowed it to be possible to create a rain-fed agricultural economy in an area in which it was rare for 300 mm of annual rainfall absent irregular to be collected.

Direct growing on pumice has created a quality agricultural landscape. Man's struggle against the scarcity of soil, aridity, and erosion has generated an agricultural area where thousands of plots are arranged in steps on the steep slopes of the area. There has also been a need to build retaining walls to modify the adverse conditions (also of pumice: stones worked by man from pyroclastic) from soil.

Farming on argillaceous soils covered with substratum of pumice is of more recent extension, although no less important from the socioeconomic point of view.

There are some sectors in the south, not covered with pumice, with argillaceous soils of a better quality which are like islets among those materials. The water-absorbing quality of pumice enables a layer of jable some 20 cm thick to make the soil preserve humidity for a long time. As a result, water consumption is noticeably reduced (1500 to 300 m^3/ha. This type of enarenado spread by reason of the water transfers taking place in the 1950s.

This system achieves a work saving: there is no need to eliminate weeds as they hardly emerge, and seeding and raking work is simple due to the low compaction of the jables. The moderating action the jable exercises on the soil because of its white color and porous nature is worthy of note.

Pumice growing on argillaceous soil currently covers more than 90% of the land farmed between 400 and 1500 m high in this area (Rodriguez Brito, 1986). Water for irrigation and the special conditions of the foreign potato market have caused the jable enarenados to become specialized in this kind of tuber. The potato monocrop is currently one of the main problems for maintaining efficiency in these agricultural systems. Pests (particularly nematodes) are causing a substantial reduction in crop productivity (this area bore two crops a year for more than a decade). Productivity can only be maintained currently on the basis of a high consumption of fertilizers and pesticides. Further, there is the high cost of water (in competition with the coastline's tourist consumption), the loss of traditional markets, and population drift towards the service sector. These are phenomena causing a large number of jable estates to be abandoned. This was not the traditional situation of the southern jables since the exchange of seeds with the north of Tenerife and between different areas in the south of varieties of local potatoes, the organic fertilizing systems, and fallow lands provided for a certain diversification of the agrosystem.

Enarenados were not the only way to improve farming. "Gavias" and "nateros" constituted — since nowadays they are practically abandoned — a way of occasional irrigation in the extensive gully network extending over the south of the island.

The most elemental form of gavia consists of building terraces at the bottom of small ravines, perpendicular to the direction of the current, so that the terraces themselves intercept runoff water. The most evolved gavias are dike built, diverting water towards land prepared on terraces on both sides of the ravine bed.

The difference between gavias and nateros lies in the fact that in the former the growing soil existed previously, while in the latter, it is necessary to build the area to be farmed (Quirantes Gonzalez, 1981).

The importance of the gavias and nateros is not due solely to using runoff water, but also because mixed with the water comes a considerable amount of sludge which is deposited on the growing area, thus contributing towards improving soil fertility.

Nateros and gavias in the south of Tenerife have been traditionally occupied by cereal and potato crops, but fruit trees are also important, e.g., fig, almond, prickly pears, and even highly water-demanding fruit trees like chestnut, pear, or apple.

CONCLUSIONS

The variety of Canary Island agrosystems is explained by the soil and climate characteristics, and in the development of a close relationship in the adaptation of agriculture to the adverse environment. The Canary island farmer has created a singular landscape of undoubted aesthetic value at the present time, but also until recently with an outstanding socioeconomic importance.

The need to protect these agricultural areas is not only because of their value as a cultural and scientific asset, but also because of the economic feasibility of the crops which have been developed here. Tourist interest in a cultural asset, scientific interest in the fight against arid land erosion, the wind and pollution, and socioeconomic interest are the result of a quality agriculture. There is also a necessary complementary relationship between agriculture and tourism.

REFERENCES

Alvarez Alonso, A. (1976). ILa organización del espacio cultivado en la comarca de Daute. Instituto de Estudios Canarios. La Laguna, 279.
Burriel de Orueta, E. (1981). Canarias: población y agricultura en una sociedad dependiente. Oikos Tau. Barcelona, 242.
Hecht, S. B. (1987). The evolution of agroecological thought. *Agroecology. The Scientific Basis of Alternative Agriculture,* M. A. Altieri (Ed.), Westview Press, Boulder, CO, 1–20.
Gliessman, S. R. (1990). *Agroecology: Researching the Ecological Basis for Sustainable Agriculture,* S. R. Gliessman, New York, 280.

Marzol Jaen, M. V. (1988). La Iluvia, un recurso natural para Canarias. Servicio de Publicaciones de la Caja General de Ahorros de Canarias. Santa Cruz de Tenerife, 220.

Quirantes Gonzalez, F. (1981). El regadio en Canarias. Interinsular Canaria, Santa Cruz de Tenerife, 2 Vols., 225 and 237.

Rodriguez Brito, W. (1986). La Agricultura de Exportación en Canarias (1940–1980). Consejeria de Agricultura, 571.

Comision Mundial del Medio Ambiente y Desarrollo: Nuestro Futuro, *Común. Alianza Editorial,* Madrid 1987.

Index

INDEX

Abandonment of land
  in Cantabrian mountains (Spain), 142
  effects on land use, 169
  in North Italy, 166
Access, as factor in Latin American agriculture and environmental classes, 201–202, 204, 209
Acer campestre, in Po Valley (Italy), 153
Acer obstusatum, in Sicily, 135
Acrocephalus arundinaceus. See Great reed warbler
Acrocephalus scirpaceus. See Reed warbler
ADDAD routines, use in analysis of farming systems, 27
Aerial photographs, use in Spanish agrarian landscape studies, 144
Afforestation
  early management operations in, 173–174
  ecological methods of, 169–176
  goals of, 169
  operation strategies for, 175
  plantation geometry in, 171
  silvicultural functionality in, 174
  site preparation for, 172–173
  spatial patterns of, 170–171
  species choice for, 171–172
Agricultural census data, use in Latin American land-use studies, 206, 208
Agricultural landscape
  above-ground insect biomass in, 71–82
  changes in Spain, 141–152
  characteristics and methods of study, 72–73
  fragmented, species survival in, 83–90
  in Po Valley (Italy), 153
  in Sicily, 131–138
Agricultural practices, effect on landscape ecology, 115, 119–120
Agricultural transport
  factors determining, 51–52
  landscape ecological patterns and, in Belgium, 49–59
Agriculture, sustainable, in Latin America, 197–214
Agrochemical abuse, in Latin America, 204, 212
Agroecological analysis
  of Latin American crops, 197
  of Latin American land use patterns, 197, 211, 212–213

Agroecological clusters, in Latin America, 197
Agroecology, definition of, 216
Agroeconomic problem studies, in Latin America, 197
Agroecosystem research
  goals of, 49
  rural road studies in, 50
  on weed ecology, 100–101
Agroecosystems
  breeding bird importance in, 153
  European, taxonomic and tropic insect groups in, 81
  landscape ecology and, 113
  land uses in, 175
Agropyron repens, in British open landscape, 16
Agrostis capillaris, in Norwegian hay meadows, 106
Agrosystems, in Canary Islands, 215–227
Alfalfa fields
  insect biomass in, 73, 74, 77, 78, 79, 80, 81
  in North Italy, 160
Allelopathic phenomena, in forest plantations, 172
All-weather roads, as factors in Latin American environmental classes, 201
Almond groves
  in Canary Islands, 226
  in Sicily, 132
Amaranthus sp., in Sicily, 133
Ampelodesmos mauritanicus communities, in Sicily, 132
Ampelographic method, application to grapevine genotype studies, 184
Ampelometric method, application to grapevine genotype studies, 184
Anchusi arvensis, in British landscape, 18
Ancient cultivars, of herbaceous plants, germplasm from, 178
Anemone nemorosa, in Norwegian hay meadows, 106
Animals
  habitat barrier effects on, 61–70
  metapopulations of, 87–88
Anthocyanin composition, of grape berry skin, use in variety studies, 184, 186, 190–192

Anthoxanthum odoratum, in Norwegian hay meadows, 106
Anthus pratensis, in North Italy, 164
Anthus spinoletta, in North Italy, 164
Anthus trivialis, in North Italy, 164
Apennine Mountains, bird population changes in, 159–160
Aphididae, 73
Apidae, in mosaic landscapes, 72
Apodemus agrarius. See Striped field mouse
Apodemus flavicollis. See Yellow-necked mouse
Apple tree cultivation
 in Canary Islands, 226
 in Sicily, 132
Apus apus, in North Italy, 163
Araneida, in unstable habitats, 36
Archipelago system model, for habitat patches, 87
Arid environments, agrosystems in, 215–227
Arrhenatheretalia, in Norwegian hay meadows, 106
Arrhenatheretea communities, in Sicily, 131
Arthropods, distribution of, 5, 37
Asphodelus microcarpus communities, in Sicily, 132
Association patterns, of rural roads, 51
Atmosphere, as link for all ecosystems, 92
Australia, weed infestations in, 95
Autecology, role in landscape ecology, 3, 5
Azores Depression, effect on Canary Islands climate, 217

Badger, in habitat patches, 87, 88
Banana cultivation, in Latin America, 208
Bank vole
 habitat barrier effects on, 63, 64, 65, 66, 67, 68
 in habitat patches, 87
Barley fields, insect biomass in, 75, 77
Barriers. See Landscape barriers
Bavaria, landscape ecology research in, 6
Bean cultivation, in Latin America, 208, 212
Belgium
 agricultural transport and landscape ecological patterns in, 49–59
 landscape ecology research in, 3, 6
Biodiversity
 from afforestation, 170
 as concept in landscape ecology, 7
Biogeography, landscape ecology and, 5

Biological reserves, in Latin America, effects on sustainable agriculture, 197
Birds
 agricultural practice effects on, 115, 119, 121, 125, 126, 127, 159–167
 breeding, in Po Valley (Italy), 153–167
 as ecological indicators, 115, 119, 121, 125, 126, 127
 fragmentation of abundance of, 162
 habitat barrier effects on, 64, 68–69
 red list of (The Netherlands), 85
Birds of prey
 landscape effects on, 99–100
 in unstable habitats, 36
Black locust, in Po Valley (Italy), 154
Brazil
 accessible area studies on, 201
 rice and cassava cultivation in, 212
 rural income per capita, 202
Brazil nut cultivation, in Latin America, 208
Brittany, water flow and hedgerows in, 43, 45, 47
Brometalia rubenti-tectori communities, in Sicily, 132
Bromus commutatis, spatial distribution of, 94
Bromus diandrus, infestations in Australia, 95
Bromus sterilis, spatial distribution of, 94
Butterflies, fragmented landscape effects on, 85

Caltha palustris, in British streamsides, 16
Canada, land-use changes in, 21, 36
Canary Island Cold Sea Current, effect on climate, 217
Canary Islands
 climate of, 216
 geography of, 216–220
 pumice cultivation on, 224–226
 traditional agrosystems in, 215–227
Cantabrian mountains (Spain), agricultural landscape changes in, 141–152
Carabid beetles, distribution of, 5, 37
Carduelis chloris, in North Italy, 163
Caribbean, environmental class mapping of, 189
Cassava cultivation, in Latin America, 208, 212
Catchment areas, as factors in Latin American environmental classes, 201
Cattle grazing
 on Canary Islands, 222, 223

INDEX

Lapilli, on Canary Island surface, 219, 221
Large-scale constraints, effects on landscape, 38
Latin America
    environmental class mapping of, 189
    land-use clusters in, 206–213
    sustainable agriculture studies on, 197–214
Leaf profiles, use in grape variety studies, 188
Lema sp., European biomass studies on, 75
Leontodon autumnalis, in Norwegian hay meadows, 106
Lepidoptera, European above-ground biomass of, 77
Leucanthemum vulgare, in Norwegian hay meadows, 106
Limestone, British vegetation on, 14
Linear barriers, effects on animal populations, 64–65
Linear features
    in British argicultural landscapes, ecological studies, 11–19
    importance of, 18–19
Linear vegetation elements, in Belgian rural road network, 52
Line transects, use in bird-population studies, 160–161
Lolium perenne, in British open landscape, 16
Luscinia luscinia. *See* Thrush nightingale
Luzula pilosa, in Norwegian hay meadows, 106
Lygeum spartum communities, in Sicily, 132

Magra watershed (Italy), bird population changes in, 159–160
Maize cultivation
    on Canary Islands, 221, 222
    in North Italy, 160
Maize fields, insect biomass in, 73
"Malpais", on Canary Island surface, 219
Manure, role in French farming practice, 124
Marginal habitats, occurrence of, 86
Maritime climates, in Latin America, 200
Marshlands, habitat characteristics of, 86
Meadows, of hay. *See* Hay meadows
Mediterranean region
    afforestation in, 173
    agricultural landscape studies on, 131–138
    bird-population changes in, 153–167
Meles meles. *See* Badger
Melolontha melolontha, in agricultural landscapes, 71
Melon cultivation, on Canary Islands, 221

Mesotrophic hay meadows, conservation of species in, 105, 106
Metapopulations
    as concept in landscape ecology, 7–8
    definition of, 83, 87
    habitat barriers within, 63, 65–65–67, 69
    spatial dynamics in, 86
    subpopulations in, 69
    survival of, 87–88
Mice, habitat barrier effects on, 63, 64, 66–67, 68
Microbes, landscape effects on, 99, 100
Microtus oeconomus. *See* Root vole
Millipedes, in habitat patches, 85
Mixed farming systems, studies in France on, 117
Modal production systems, relation to Latin American environmental classes, 208
Molino-Arrhenatheretea, in Norwegian hay meadows, 106
Montane circum-Mediterranean region, bird-population changes in, 163
Morus alba, in Po Valley (Italy), 132, 156
Mosaic landscapes
    above-ground insect biomass in, 71–72, 74, 76, 77–78, 81
    characteristics of, 72
    effects on Italian bird populations, 159, 160, 163, 166
    wild animals and plants in, 85
Mosses, in Norwegian hay meadows, 106, 110
Mountain agriculture, in Spain, 141–142
Mulberry, in Po Valley (Italy), 132, 156
Mulching, use in afforestation projects, 174
Multiple correlation analysis, of grape varieties, 184, 185
Mutual Information Index, use in Spanish agrarian landscape studies, 145

"Nateros", role in Canary Island agriculture, 226
National parks, in Latin America, effects on sustainable agriculture, 197, 201
Natural resource management, in Latin America, 199
The Netherlands
    habitat-patch studies in, 87, 88
    landscape ecology research in, 3, 6
    red list of plants in, 85
    rural road planning and studies in, 50
Network of hedges and hedgerows, in France, 41–42
New Zealand, conifer plantations in, 171–172